NETWORKED OPERATIONS
AND TRANSFORMATION

Networked Operations and Transformation

Context and Canadian Contributions

ALLAN ENGLISH, RICHARD GIMBLETT,
HOWARD G. COOMBS

McGill-Queen's University Press
Montreal & Kingston • London • Ithaca

ISBN 978-0-7735-3285-4 (cloth)
ISBN 978-0-7735-3339-4 (paper)

Legal deposit fourth quarter 2007
Bibliothèque nationale du Québec

Printed in Canada on acid-free paper that is 100% ancient forest free (100% post-consumer recycled), processed chlorine free

This study was prepared for the Canadian Department of National Defence but the views expressed in it are solely those of the authors. They do not necessarily reflect the policy or the opinion of any agency, including the Government of Canada and the Canadian Department of National Defence.

McGill-Queen's University Press acknowledges the support of the Canada Council for the Arts for our publishing program. We also acknowledge the financial support of the Government of Canada through the Book Publishing Industry Development Program (BPIDP) for our publishing activities.

Library and Archives Canada Cataloguing in Publication

English, Allan D. (Allan Douglas), 1949–
    Networked operations and transformation: context and Canadian contributions / Allan English, Richard Gimblett, Howard Coombs.

Includes bibliographical references and index.
ISBN 978-0-7735-3285-4 (bnd)
ISBN 978-0-7735-3339-4 (pbk)

1. Western countries – Armed Forces – Technological innovations.
2. Western countries – Armed Forces – Reorganization.
3. Command and control systems – Western countries. 4. Military art and science – Technological innovations. I. Gimblett, Richard Howard, 1956– II. Coombs, Howard III. Title.

U42.E54 2007 355.4'71713 C2007-902618-4

Typeset by Jay Tee Graphics Ltd. in 10.5/13 Sabon

Contents

Abbreviations

2GW	Second Generation Warfare
3D	Defence, Diplomacy, and Development
4GW	Fourth Generation Warfare
AAW	anti-air warfare
ADF	Australian Defence Forces
ADSI	Automated Data Systems Integrator
ANA	Afghanistan National Army
ASUW	anti-surface warfare
ASW	anti-submarine warfare
ATA	Afghanistan Transitional Authority
AUS	Australia
AUSCANUKUS	Australia and Canada and the United Kingdom
AWACS	Airborne Warning and Control System
BFEM	Battle Force E-mail
C^2	command and control
C4I	Command, Control, Communications, Computers, Intelligence
C4ISR	Command, Control, Communications, Computers, Intelligence, Surveillance, and Reconnaissance
C@S	Collaboration at Sea
C@S2	Collaboration at Sea, Version 2
CCC	Civilian Conservation Corps
CDN	Canadian
CENTCOM	US Central Command
CENTRIXS	Combined Enterprise Regional Information Exchange System
CF	Canadian Forces

CFC-A	Coalition Forces Command – Afghanistan
COWAN	Coalition Wide Area Network
CTF	Commander Task Force
CVBG	carrier battle group
CWC	Composite Warfare Commander
DAD	Directorate of Army Doctrine (Canada)
DCDS	Deputy Chief of the Defence Staff
DDH	helicopter-carrying destroyer
DND	Department of National Defence
DRDC	Defence Research and Development Canada
EBO	Effects-Based Operations
GCCS	Global Command and Control System
GCCS-M	Global Command and Control System, Maritime
GWG	Global War Game
HF	high frequency
HUMINT	human intelligence
IMD	Information Management Director
IMF	Investment Management Framework
ISAF	International Security Assistance Force
ISTAR	Intelligence, Surveillance, Target Acquisition, and Reconnaissance
JIMP	Joint, Interagency, Multinational, and Public
JOTS	Joint Operational Tactical System
LAN	Local Area Network
LFC^{2}IS	Land Force Command and Control Information System
MCOIN	Maritime Command Operational Information Network
MIO	maritime interdiction operations
MNFC	Multinational Force Commander
NATO	North Atlantic Treaty Organization
NAVCENT	US Naval Forces Central Command and commander, US Fifth Fleet
NCM	non-commissioned member
NCW	Network-Centric Warfare
NDHQ	National Defence Headquarters
NEC	Network-Enabled Capabilities
NEOps	Network-Enabled Operations
NORAD	North American Aerospace Defence
NPPS	National Priority Programs

OODA	observation-orientation-decision-action or Observe, Orient, Decide, Act
OPTASK	operational tasking
OTC	Officer in Tactical Command
PACOM	US Pacific Command
RDO	Rapid Decisive Operations
RMA	Revolution in Military Affairs
RN	Royal Navy
ROE	rules of engagement
SAMS	School of Advanced Military Studies (US Army)
SATCOM	satellite communications
SHIPLAN	ship local area network
SIPRNET	Secret Internet Protocol Router Network
SME	subject matter experts
SNFL	Standing Naval Force Atlantic
SOF	special operations forces
SOSUS	Sound Underwater Surveillance System
SPDA	Structured Process for the Harmonized Development of Afghanistan
Spec Ops	special operations
STOM	ship-to-objective manoeuvre
TACNOTE	tactical procedure note
TACOM	tactical command
TACON	tactical control
TADIL	tactical datalink
TF	Task Force
TRUMP	Tribal Update and Modernization Program
UAV	unmanned aerial vehicle
UHF	ultra-high frequency
UN	United Nations
UNAMA	UN Assistance Mission in Afghanistan
UNEF	United Nations Emergency Force
USN	US Navy
USW	Under-Sea Warfare
weCAN	Web-Centric Anti-Submarine Warfare Net
Y2K	Year 2000

Foreword

Networked operations, which are based on the concept of Network-Centric Warfare (NCW), are at the heart of many ongoing transformation initiatives in Western militaries. While terms like NCW are frequently used in discussing both how operations should be conducted in the current and future security environments as well as how Western armed forces should be transformed, these terms are used by different people to mean different things. These differences in meaning result from the confusion in many circles as to what the concept of NCW actually entails. Therefore, it seems timely to address the concept of networked operations, especially as it relates to transformation, by focusing on its origins and how it relates to other concepts, like the Revolution in Military Affairs, operational art, manoeuvre warfare, Rapid Decisive Operations, and Effects-Based Operations. Without a clear idea of the context of the networked operations concept, decision-makers and military professionals may not understand the implications of their decisions and actions in planning and implementing transformation initiatives.

The genesis of this book was a meeting called by Carol McCann at Defence Research and Development Canada (DRDC) – Toronto in March 2005 to discuss the implications of the concept of networked operations, particularly NCW, for Canadian Forces (CF) operations in light of the increasing prominence of a Canadian version of networked operations – Network-Enabled Operations, or NEOps. Those attending the meeting were military officers, academics, and retired officers who were now in the academic community, all of whom had detailed knowledge of some aspects of networked operations.

Two of the authors of this book, Allan English and Howard Coombs, participated in the meeting, and it was apparent to us that the other participants were using terms like Effects-Based Operations and NCW in different ways. This demonstrated that there were different interpretations of these and other terms that came under the networked operations umbrella. When asked by some defence scientists whether there was a theory of NCW to give some coherence to the various interpretations of networked operations, we replied that there wasn't, but that it would be worth documenting different interpretations of the terms in use so that people studying the issue would be aware of some of the pitfalls of unconditionally accepting networked operations as one of the principal concepts of CF transformation.

This book began as a study of networked operations for DRDC, and we quickly realized that besides joint interpretations of networked operations, we also needed to describe army, navy, and air force perspectives. Although we were comfortable with exploring the joint, army, and air force perspectives, we needed an expert in naval operations, especially Canadian naval operations, to explore naval perspectives. Therefore, we engaged our colleague and naval analyst Richard Gimblett in this study, who also traced some of the Canadian roots of NCW.

The result is this study, which aims to contribute to the literature on networked operations, particularly as it is being applied to transformation in armed forces in the Western world. The concept of networked operations is likely to continue to be influential in the security and defence communities. However, we believe that the theoretical and practical aspects of networked operations must be examined carefully and understood clearly before this concept is accepted unreservedly as the driving concept behind transformation.

Allan English, Richard Gimblett, Howard Coombs

NETWORKED OPERATIONS
AND TRANSFORMATION

1

Introduction

The difference between a tradesperson, or a technician, and a professional, is that the former uses tools without a comprehensive understanding of how these tools came to be, and why, whereas such knowledge is the key to the latter's ability to adapt those tools for use in new and innovative ways or to modify them to meet unforeseen requirements.

David Bercuson[1]

At the beginning of the twenty-first century the concept of networked operations has come to the fore of ways of thinking about warfare and other operations involving military and security forces. The leading approach to networked operations is called Network-Centric Warfare (NCW). This concept was originally developed by the US Navy, but it now dominates US military transformation initiatives and is being used as a template for future American command and control (c^2) frameworks. Networked operations are currently touted as the way to fundamentally change how the US and, by extension, coalition forces will conduct operations.

There is still some confusion, however, as to what the concept of NCW actually entails, both because definitions of NCW have changed since it was first formally introduced as a concept over seven years ago[2] and because of what some critics have referred to as its "jargon-laden language."[3] In the late 1990s NCW was seen largely as a fully integrated information network, with all platforms being nodes in the network. The primary aim was to produce a "common operating picture" so that all players would be working from the same computer-mediated visual presentation.[4] Today, its principal architects describe NCW as two things: "an emerging theory of war in the Information Age," and "a concept that, at the highest level, constitutes the military's response to the Information Age." We are told in the latest official US policy statement, *The Implementation of Network-Centric Warfare*, that NCW is now about "the combination of strategies, emerging tactics, techniques, and procedures, and organizations" that networked forces can use to "create a decisive warfighting advantage."[5] Despite these declarations, Canadian researchers have noted the lack of a succinct definition of NCW in official publications.[6]

A concrete example of one of the latest iterations of NCW is the US Navy and US Marine Corps' "functional concept" for future operations – FORCEnet. As one of the latest derivations of NCW and one closest to its US Navy roots, FORCEnet gives us some insights into the most recent concepts underlying networked operations. A paper issued by the US Department of the Navy "to establish a common direction for the diverse efforts that contribute to building naval command and control capabilities in the future and to provide a common framework for thinking about future command and control"[7] says: "The foundation of FORCEnet is a *fully integrated, self-healing, self-organizing* communications system or infrastructure ... To optimize network effects, the infrastructure will be based on a *modular, open-systems architecture* which allow all nodes to interact regardless of location or network address [emphasis in original]."[8]

Although the FORCEnet concept acknowledges the human dimension of networked operations, this acknowledgment belies the technology-centric approach found in most of the derivatives of NCW: the FORCEnet paper states that in the design of this new approach to warfare and other operations, "technology solutions are often the most obvious" and that technology should "co-evolve with the other elements of force development." In line with this technology focus, three of the six dimensions of FORCEnet are physical, and only one is explicitly human-centred.[9] This technology focus may be putting the cart before the horse in ways that militaries have done in the past: buying the newest and the best equipment without any clear idea how it might affect, negatively as well as positively, the way in which operations are conducted. From a Canadian perspective, two main deficiencies stand out in current NCW concepts: their emphasis on war fighting versus other types of operations, and their focus on technology over the human dimension in conducting operations.

Like other concepts that have had a major impact on how Western militaries think about warfare and other operations in the past twenty years (e.g., the Revolution in Military Affairs [RMA], operational art, manoeuvre warfare, Rapid Decisive Operations, and Effects-Based Operations), networked operations have their origins in the US and therefore have had a major impact on how other Western nations think about military and security operations. But it would be imprudent to fully embrace NCW-based concepts because, whatever their strengths and weaknesses, there is no guarantee they will endure. NCW exaggerates certain fashionable technological features of ideas about war fighting,[10] and like other concepts that have catered to the latest fads, networked operations now has its turn to have a place in the sun. Yet, because of its genesis in late twentieth-century naval warfare operations and business practices, reinforced by experience in certain post-Cold War military campaigns (like Desert Storm), the sun may already be setting on networked operations as "post-hostilities" campaigns in Afghanistan and Iraq challenge some of its basic tenets.

Some theories stand the test of time, while others do not apply very well across temporal and cultural boundaries. Whether NCW will shortly become passé, will depend on a number of factors. Some of these factors include the theoretical foundations of NCW and the cultural implications of adopting it as an overarching principle of transformation. Yet there is very little discussion about its origins and, perhaps more importantly, how various theoretical concepts have been lumped together, under different names, to describe varied visions of networked operations. For example, a recent paper from the Canadian Forces' office of the Director General Joint Force Development refers to "the theory of Transformation and the associated concepts of Effects Based and Network Centric Operations," when in fact there are a number of "theories" of transformation, and Effects-Based Operations and network-centric operations are not closely "associated" either with each other or with transformation, as we shall see.[11] There is a danger in this approach of combining distinct theoretical constructs into one amalgam: as one commentator on operational art noted in the formative years of that concept, attempting to combine too many diverse, and potentially incompatible, ideas under one umbrella term "may invite only muddle."[12]

This muddle is apparent in the very nomenclature associated with networked operations today. The dominant strand in networked operations discourse originates from the American concept of Network-Centric Warfare, but other countries have taken NCW and adapted and modified the concept to suit their own needs and military cultures. For example, the Australian military has put more emphasis on the human dimension of NCW than was found in the original US model. Other countries have gone further in their adaptations and created new names for their brand of NCW. For example, the UK now uses the term Network-Enabled Capabilities (NEC), which is described as more "'commander-centric'" than "'network centric.'" Canada has adopted the term Network-Enabled Operations (NEOps), an evolving concept that is to be linked to other Canadian initiatives like the "3D" (defence, diplomacy, and development) approach to security.[13] However, as with NCW, there is no officially approved, concise definition of NEOps.[14] Arguably, the best succinct definition of NEOps to date is "the conduct of military operations characterized by common intent, decentralized empowerment and shared information, enabled by appropriate culture, technology and practices."[15] The problem with some of the efforts to develop the NEOps concept in Canada is a lack of awareness of the assumptions and cultural outlooks that are imported with NCW and other approaches to networked operations. In developing new concepts it is useful to borrow and synthesize good ideas from others; however, it is essential to have a solid grasp of the underlying values and beliefs that come with these concepts.

Therefore, this volume examines theoretical and historical origins of the concept of NCW and related networked operations to more fully understand the nature of Networked-Enabled Operations today and how it might evolve in the future. We conclude that, given the nature of NCW, and its origins and underpinning assumptions, Canada and other Western nations should be careful about completely embracing NCW as the basis for their approaches to warfare and other operations. This caution extends particularly to nations like Canada that cannot afford all the technology to implement a fully developed NCW architecture. Furthermore, as we shall see, Canada has a long and successful history of putting human networks ahead of technical networks – in our view the proper approach when transforming armed forces to meet future challenges. Therefore, while it is necessary to borrow some concepts from other approaches and to devise some new concepts through experimentation, we argue here that Canadian military professionals and others should draw on Canada's extensive experience with human-centred networks, leveraged by a judicious use of select technologies, in creating our own approach to NEOps.

THE MILITARY AS A LEARNING PROFESSION

As Bercuson maintains above, the military profession, like other professions, uses theories to explain why and how things work. Just like professional engineers, who study both the theoretical and practical aspects of their discipline, military professionals must understand the theories of their field of practice before they can be called true professionals. Furthermore, in today's world of transforming militaries, where NCW or NEOps is on many transformation agendas, leading US military educators have asserted that the successful transformation of military organizations requires "intellectual leaders" who, among other things, are educated as professionals in military art and science, including theories of war understood within their historical context.[16]

The notion of networked operations can thus be seen as one of a number of tools to be used by military professionals and others. Therefore, those who expect to design or to work with the NEOps tool must have a comprehensive understanding of how and why this tool was created and how and why it is changing shape if they wish to adapt it in the future. This book is a preliminary step in understanding the origins and evolution of the networked operations tool so that interested professionals can adapt it for use in new and innovative ways or modify it to meet unforeseen requirements.

The historical context of NEOps includes the sources of the theories behind its development. The background and character of the theorists who codified our ideas on war, including ideas related to NCW and NEOps, are important because these theorists have been shaped by a number of influences,[17] such as their personal experience, ideology, religion, culture, and economic circumstances (i.e., who pays them).

Some military officers, and some defence department bureaucrats in the West, still do not accept the need to study theories of war and history as part of their professional development or as key components of the practice of their profession. They hold a view that abstract theories and historical experience are not useful in the real world of war. Recently, these views have gained greater acceptance in some circles owing to the high operational tempo of the post-9/11 "war on terror," which has led to cutbacks in senior officer and other professional military education in some Western nations. These cutbacks have been critiqued by a leading US defence intellectual:

> Recently, one defense official defended a proposal to shut down temporarily parts of the Army's advanced professional military educational system with the remark, "Some of the experiences they are getting today are better than anything they will get in a classroom ... It's not giving up something for nothing. We have a generation of leaders in the Army today that are battle-tested and are much more capable of leading the Army from the actual experience they have."
>
> The stupidity of this last remark is ... depressing ... It implies that knowing how to maneuver a battalion through an urban fight is the same thing as crafting a strategy for winning a counterinsurgency. It suggests that at least some at the top of the Pentagon do not understand that the next war will be as different from Iraq 2005 as Iraq was from Somalia, and Somalia from Panama, and Panama from Vietnam. Combat experience can indeed give us an army that can fight and win America's battles; but it is education that provides the intellectual depth and breadth that allows soldiers to understand and succeed in America's wars ... some of our most successful commanders in Iraq declare that their master's and Ph.D. degrees in history, or political science, or anthropology, provided some of the best preparation possible for the novel challenges of insurgent warfare. Senior military leaders, and a few civilians, acknowledge the existence of the problem but seem to lack the ability or the will to do something about it.[18]

Similarly, because the debate over networked operations is based on abstract concepts (e.g., synchronization, integration, manoeuvre, levels of war, etc.), some degree of advanced professional military education is required to master them. Therefore, it behooves military professionals to study their profession so that they might do it well and, like a skilful surgeon, minimize the effects of necessary surgery on the body politic. Finally, theories are an important part of the process of creating doctrine. Along with an analysis of technological advances and recent historical experience, theory is one of the key ingredients to developing effective doctrine.[19] Therefore, theory and history are essential adjuncts to military professionals in the study and practice of their profession.

THEORETICAL FOUNDATIONS

Advocates of NCW assert that it is an emerging theory of war that is based on the Tofflers' waves theory of warfare and the notion that we now live in the information age, where "third wave" high-technology information warfare will become the new standard for success in war fighting and other operations.[20] However, critics have challenged these and other assumptions underpinning NCW. Therefore, it is necessary to briefly examine the nature of theories of war.

Related to this question is the debate in parts of the literature on whether networked operations belongs primarily in the domain of military art or military science. We argue here that networked operations is a multidisciplinary topic that requires insights from many academic disciplines as well as from practitioners; therefore, debate over the proportion of art and science in this concept is largely sterile. To illustrate this point, we know that successful creative artists like painters must have, besides artistic talent, an understanding of sciences like geometry (to understand concepts like perspective) and chemistry (to work with various media, like paints). Likewise, the most successful scientists acknowledge a creative or artistic side to their work. For example, Canada's second Nobel laureate in physics, Bertram Brockhouse, was described by a colleague as having "this gift of just knowing what the answer is, then doing the experiment to prove it."[21] This statement bears an uncanny resemblance to an observation by a critic of NCW on the methodology used by those charged with developing this concept: "This is a relatively new idea and the theory calls for extensive experimentation. But the way it is being done implies they already know the outcome of the experiments."[22] For those who believe that the outcome of experiments should not be predetermined, the methodology being used to develop NCW is flawed.

The debate over art versus science notwithstanding, some concepts need to be clearly understood because they are important in comprehending the context in which tools like NCW and NEOps originated and are evolving. Two of these concepts are theories and paradigms.

NCW AS A THEORY The authors of the most recent articulation of the NWC concept, *The Implementation of Network-Centric Warfare*, have selected definitions of the word "theory" that emphasize its speculative nature in some common usage.[23] This usage of theory, which is defined in one dictionary as "a speculative or conjectural view of something," accords closely to the understanding of Prussian author and soldier Carl von Clausewitz (1780–1831) as to how theories should be used. Clausewitz's approach was strongly influenced by Kantian philosophy, and he used the dialectic approach of thesis, antithesis, and synthesis to study the subject of war. In his book *On War* he constantly revises his hypotheses and moves back and forth between the ideal and the real states of war.[24] Many of the writings found in the American professional military literature, however, quote Clausewitz out of context as if he had written a book of instruction on the conduct of war. But he did not; he wrote a treatise to help us better understand the phenomenon of war through debate and the synthesis of competing concepts.[25]

The implications of the speculative approach to theories have important implications for NCW theory. It could be argued that NCW "theory" is no more than a series of largely untested hypotheses or assumptions that should be subjected to research and a Clausewitzian dialectic to determine their usefulness. While this approach is nothing new in the history of military theories, it has profound implications when one observes how completely NCW has been embraced as the benchmark for US Department of Defense transformation and, by extension, some other Western nations' military transformation. In many policy documents NCW is often portrayed not as a speculative theory but as an authoritative doctrine on future warfare. Accepting it as such and embracing it completely may be a high-risk activity because the transformation of militaries and their future may be based on untested speculation. This problem is exacerbated for nations like Canada that are creating their own "theories" of networked operations on what may be the proverbial foundations of sand that are washed away by the next "wave" in military theory.

A number of organizations have adopted NCW as though it were an integrated view of the fundamental principles underlying a science or its practical applications.[26] In this view of a theory, if its principles are correctly applied, the theory is generally accepted to have explanatory power. This description might be used to characterize the approach of Swiss-French author and soldier Baron Antoine-Henri Jomini (1779–1869) to the study of war. Jomini emphasized decision-making rules, operational results, and the conceptualization of warfare as a huge game of chess. His conception of war has been surprisingly durable in the present age of computer-mediated warfare, where the Jominian paradigm underpins much of the Western approach to modern warfare.[27] In today's world, where our lives are strongly influenced by scientific notions, we usually expect a theory to be able to explain causality or why things happen.[28] Therefore, when many military professionals see the word "theory" attached to a concept, they expect it to have considerable explanatory power.

The latest policy statement on NCW, *The Implementation of Network-Centric Warfare*, offers an interesting paradox when it attempts to combine aspects of both Clausewitz's and Jomini's approaches to its own theoretical approach. On the one hand, it concludes that "classic strategic theories of war may require adaptation to a changing environment ... [but that] they remain fundamentally intact. The logic of waging war and of strategic thinking is as universal and timeless as human nature itself." Furthermore, this policy statement acknowledges that a large number of theorists at the end of the twentieth century proposed a number of alternative frameworks for war in the future. On the other hand, by adopting the concept of the information age as its foundation, *The Implementation of Network-Centric Warfare* has not attempted a synthesis of previous theories of war but has pinned its hopes on one specific interpretation of war. The Tofflers' interpretation of war occurring in waves, with the current wave being based on information, has been challenged by a number of commentators, including Steven Metz, who is currently teaching at the US Army War College's Strategic Studies Institute. He argues that "Quintessentially American, the Tofflers concentrate on technology feasibility with little concern for the strategic, political, social, psychological or even ethical implications of changing military technology." He states that their theories are particularly attractive to the US military because of their relatively simple, if flawed, interpretation of war.[29] Despite a recognition of the importance of the human in the latest NCW policy documents, it is important to remember that NCW theory is founded on an essentially technological approach to war.

Another problem with applying NCW theory is based on US military culture. *The Implementation of Network-Centric Warfare* recognizes that the theoretical constructs of the classic theorists of war "remain fundamentally intact," and among the classical theorists Clausewitz is cited most frequently as the basis for the doctrinal writings of the US services. However, his theories are in some respects at odds with the assumptions underlying NCW, which may cause difficulty in adapting NCW to current or future doctrine.[30] Furthermore, as Paul Johnston has demonstrated, the characteristics, historical experience, and culture of an armed force may have an important impact on how it plans to fight and how it actually performs on the battlefield.[31] If implementing NCW requires major cultural changes in armed forces, its advocates should take into account the fact that successful cultural change often takes a considerable amount of time; such change is usually measured in years, and even decades, because major culture change may require paradigm shifts in the organization. For example, in an attempt to deal with operations in the "new world disorder," some critics have noted that "the Pentagon as an institution appears ... unable to shift from a network-centric warfare to a culture-centric warfare paradigm."[32]

PARADIGM SHIFTS The process of paradigm shift often has significant effects on how a particular theory affects an organization's predominant paradigm. In some cases the new theory effects a paradigm shift; in other cases the theory is rejected or modified to fit into the prevailing paradigm.[33] Azar Gat, a leading writer on military thought, notes that "[n]ew and significant intellectual constructions usually emerge at times of fundamental change or paradigmatic shifts, when prevailing ways of interpreting and coping with reality no longer seem adequate."[34] But during paradigm shifts, new notions and concepts are often hazy and ill defined. Part of the reason that NCW concepts are unclear is that NCW was developed at the end of the Cold War, when a paradigm shift about the nature of war and conflict was underway. And yet NCW drew on theories whose antecedents came from the Cold War period, particularly naval operations and certain aspects of business theory.

Therefore, in examining the history of NCW, this volume focuses on its origins, because its roots in naval operations and business theory have nurtured certain assumptions that are found in the fruit of the NCW theoretical tree. These assumptions are not compatible with all military cultures, and may actually be antithetical to some. Each country's military and each service within that military has its own war-fighting paradigm based on the physical environment in which it fights as well as its technology, history, and culture. Some of these paradigms can accept the notions embedded in NCW easily, some can accept these notions but must modify them to work in a different framework, and others are not compatible with the key assumptions of NCW.

Throughout this discussion, it would be wise to keep in mind Williamson Murray's thoughts on theories and models. He argues that while they can aid analysis, they can offer no formulas for the successful conduct of war, because its reality is far too subtle and complex to be encompassed by theory. At best, he claims, theories can provide a way of organizing the complexities of the real world for studying war because, as Clausewitz suggests, "principles, rules, even systems" of strategy must fall short in a domain where chance, uncertainty, and ambiguity dominate. And yet, while many variables that cause ambiguity have effects that differ from one situation to another, others have effects whose features recur with impressive regularity.[35]

2

Frameworks for Thinking about Networked Operations

Western navies, air forces, and armies have adopted certain paradigms and theories of war, and it is important to understand the context in which these paradigms and theories reside to understand how NCW may or may not apply to navies, air forces, and armies. At the end of the twentieth century a number of ways of thinking about conflict, warfare, and military operations emerged and dominated discussions about future war in the Western world. The first part of this chapter briefly examines two major ways of thinking to put the discussion on NCW and NEOps that follows into context. The second part briefly describes two ways of thinking about command, control, and c^2, issues that are at the heart of NCW.

WAYS OF THINKING ABOUT FUTURE WAR AND CONFLICT

THE REVOLUTION IN MILITARY AFFAIRS In the last decade of the twentieth century the term "Revolution in Military Affairs" was the catchphrase that symbolized how change would affect future war. More recently, terms like "transformation" and NCW have begun to replace RMA in dialogues about the future of war.

In the 1990s, the idea that we were in the midst of an RMA that would transform war was endorsed by the senior leadership of the CF in an effort to maintain the CF's interoperability with US armed forces and other Western militaries that were espousing this concept.[1] The US Office of Net Assessment defined an RMA as "a major change in the nature of warfare brought about by the innovative application of new technologies which, combined with dramatic changes in military doctrine and operational and organizational concepts, fundamentally alter the character and conduct of military operations."[2] For some, NCW is a direct descendent of the RMA concept, and they point out that over a decade ago devotees of the RMA made promises about fundamentally changing military operations similar to those made by advocates of NCW today.

While there is no doubt that the ideas and technology spawned under the banner of the RMA changed warfare, there is still significant disagreement over whether the changes were revolutionary or evolutionary. Some have suggested that if we look around the globe today we see nothing unprecedented in human conflict, since nationalist, religious, and ethnic conflicts are hardly particular to the late twentieth or early twenty-first century. These types of war have existed since at least the Middle Ages and exhibit significant continuity with the evolution of warfare over the past 100 years. So, according to some, if we are in the midst of an RMA we can make a strong case that it has been unfolding for a long time.[3] This perspective concerning the RMA is best articulated by American military historian Robert Bauman, who suggested in 1997 that while new challenges will arise as a result of emergent discoveries, the fundamental nature of warfare will not change. Bauman argued that challenges in "linking futuristic technological change and doctrinal concepts are much the same as they were a century ago" and that many of the current trends exhibit threads of continuity in the evolution of conflict since the nineteenth century.[4] Therefore, he believes that there is not enough evidence to support the notion that the RMA represented a paradigm shift. The dangers of focusing too much on the technical determinants of change in warfare with concepts like the RMA have been discussed by a number of commentators, and recent events in Iraq have reminded us that some problems encountered in war and conflict are still not susceptible to technical solutions, as we shall see later.

TRANSFORMATION "Transformation" is the latest buzzword used to symbolize how change will affect future war, and NCW is at the heart of current US Department of Defense transformation efforts. The most recent approach to transformation in Western armed forces began when then US Secretary of Defense Donald Rumsfeld established the Office of Force Transformation in October 2001 and gave it the mission of coordinating all the US services' transformation efforts. The Office's first director, the late Admiral Arthur K. Cebrowski (US Navy retired), put his imprint on US transformation efforts; therefore, a brief look at his background will help us to understand the concept. Cebrowski, a naval aviator, had thirty-seven years of service in the US Navy that included combat experience in Vietnam and Desert Storm, command of a carrier battle group, and president of the US Naval War College. Frederick Kagan characterized Cebrowski's new vision as resonating with his operational experience as a naval officer who was used to operating in a fluid medium against a relatively limited arrays of targets. Perhaps most importantly, Cebrowski was instrumental in developing and publicizing NCW as a distinct vision of future warfare. It should come as no surprise, then, that Cebrowski "enshrined NCW as the goal" of transformation and declared that the "transformation programs in the services will be judged by the extent to which they approach the NCW ideal."[5] However, there are problems with NCW, as alluded to above, that have been pointed out in the literature practically since its inception, and especially in the past five years.[6]

One of the biggest problems in trying to understand and apply the terms most frequently used in the debate about NCW and the future of war is that there are a number of different interpretations of these terms, and there is, as yet, no overarching theory to link them together. A reading of *The Implementation of Network-Centric Warfare* and "FORCEnet: A Functional Concept for the 21st Century" reveals that a number of distinct approaches to war and other operations, (e.g., manoeuvre, Effects-Based Operations, elements of the OODA loop, and information age warfare) are all gathered together under the umbrella of networked operations. Unfortunately, some of these modern theories of war are not always well articulated or described, as we shall see, and yet they are integral parts of the evolving NCW concept.[7] Both documents also ignore the differences and inconsistencies among these concepts and blithely assume that they can all be combined into a new theory of war – NCW. It may be possible to do this; however, it may also be the case that theories developed by different individuals and organizations from culturally diverse backgrounds are not compatible.

Critics of NCW and of current US transformation efforts highlight the terminological confusion in NCW documentation by pointing out that even "transformation" is not clearly defined by those in charge of these efforts. As Andrew Krepinevich Jr noted in 2004, "One of the problems with the transformation effort is that, three years into it, there is not a clear understanding at the Pentagon of what the term means ... It's become more a generic buzzword for ill-focused change."[8] Cebrowski himself provided no clear definition of transformation. Thomas Ricks and Josh White, writing in the *Washington Post*, quoted Cebrowski as saying this about transformation on the Office of Force Transformation Pentagon website: "'Some say it is about injecting new technology into the military ... Others believe transformation is about new ways of buying weapon systems. Still others hold that transformation is about the wholesale change of organizations ... Frankly, I don't care which one is used,' as long as it is understood to be a process that keeps the U.S. military changing and competitive in warfare." They further reported, again quoting Cebrowski, "there was a good reason not to dwell on what exactly is meant by transformation: 'I've watched senior leaders get knotted up in the definition of transformation' and lose their focus on substance ... His bottom line, he said, is that 'what we're really talking about is changing behavior.'"[9]

However, changing behaviour may not be enough. The real problem, as Elinor Sloan has observed, is that the "dominant military service cultures" continue to focus on legacy equipment, as reflected in procurement budgets that do not fully reflect service visions.[10] This observation has been reiterated by Krepinevich, who was quoted as saying, "'There are efforts in transformation in some areas – like UAVs [unmanned aerial vehicles] and networked Navy battle groups – but if you look at the overall budget, what you see are the legacy programs' ... Most of the spending, [Krepinevich] said, goes to large ships, submarines, fighter aircraft and other programs that he calls 'the traditional force structure items,'"[11] as efforts to modernize the services' major budgetary decisions have essentially failed. Some of the reasons behind the resistance to change can be found in the dominant military service culture, as Johnston noted.

Finally, this confusion in NCW terminology is compounded by the fact that the proponents of NCW have selected mainly a top-down approach to transforming the US armed forces. While analyzing this type of approach is beyond the scope of this work, it should be noted that it runs the risk of losing momentum once its key support-ers leave their posts and are unable to provide continued impetus to the initiative.[12]

WAYS OF THINKING ABOUT COMMAND, CONTROL, AND C^2

The terms command, control, and C^2 are used repeatedly in NCW policy documents, as these issues are at the heart of NCW. And yet, as is stated in CF leadership doctrine, it is sometimes difficult to dis-tinguish among these terms, and related terms like management, because the complexity and "inter-relationships and interconnec-tedness of command, management, and leadership *functions* often make it difficult to disentangle the command, management, and leadership effects achieved by individuals in positions of authority. Hence favourable results tend to be attributed to extraordinary leadership even when they may, in fact, be the result of command or management skills, some combination of all three, or other factors – including luck [emphasis in original]."[13] Nevertheless, in this part of the chapter two leading frameworks are described that provide ways of thinking about these issues that can help to clarify their use in NCW and NEOps.

The theoretical study of command in a military context is still immature, and terms like "inchoate," "diffuse," "conjectural," and "seemingly random" have been used to describe current approaches to this subject. Furthermore, in practice, there is confusion among some branches of the military in terms of how they describe their approach to command. Some endorse the concept of mission com-mand, others endorse a philosophy of centralized control and decentralized execution, while in other services the notion of net-work-centric C^2 is prominent.[14] This part of the chapter gives an overview of some of the main theoretical and practical issues related to command at the operational level. In an attempt to put some order into the discussion, this overview relies on two frame-works: the first devised by Thomas J. Czerwinski, who served in the US Marine Corps and US Army and was on the faculty of the National Defense University, and the second put forward by Cana-dian researchers Ross Pigeau and Carol McCann of Defence Research and Development Canada (DRDC) – Toronto. Both frame-works are referred to in the literature on NCW, and the latter framework is emphasized here because it is one of the leading empirically based models of c^2 currently being developed.

THE CZERWINSKI FRAMEWORK Czerwinski proposes a framework that summarizes many of the concepts in the current debate and is based on three types of command style. He describes the first command style, used in the US Army's Force XXI/digitized battlefield concept, as "command-by-direction." This form of command has been used since the beginning of organized warfare, and it is based on commanders attempting to direct all of their forces all of the time. A common example of command-by-direction is Napoleon or Wellington sitting on horseback overseeing a battle and issuing commands by sending mounted aides to pass on their orders to subordinate commanders. This form of command became progressively more difficult to exercise from the middle of the eighteenth century onwards as the size of forces in the field increased to the point where they could no longer all be seen by the commander. Czerwinski argues that command-by-direction has been resurrected by the US Army because it believes that technology can provide the commander with the ability to exercise this type of command again, using electronic displays to "see" and new communications systems to transmit orders to all of the commander's forces. However, he asserts that because of the size and complexity of the technical support required to support this command style, it will be inadequate and self-defeating if applied to twenty-first-century conflict.

Czerwinski's second style, "command-by-plan," was created by Frederick the Great 250 years ago to overcome the limitations of command-by-direction. Command-by-plan emphasizes adherence to a predetermined design, and it has evolved as the norm for many modern military forces in the West. The US Air Force's air campaign doctrine is cited as an example of this type of command system, in which focus replaces flexibility so that an opponent's centres of gravity can be identified and neutralized. At the heart of this doctrine is a detailed campaign plan implemented through a series of daily Air Tasking Orders that specify in minute detail how each aircraft's mission is to be conducted to achieve the aims of the campaign plan. Czerwinski claims that command-by-plan is useful only at the strategic and operational levels of war, and if too much emphasis is put on adhering to the plan, this method will be ineffective because of its inability to cope with unforeseen or rapid change.

Czerwinski advocates the adoption of a third type of style, "command-by-influence," also called mission command, to deal with the chaos of war and the complexity of modern operations. This command style attempts to deal with uncertainty by moving decision thresholds to lower command levels, which thereby allows smaller units to carry out missions bounded by the concept of operations derived from the commander's intent. The emphasis in this method of command is on training and educating troops to have the ability to exercise initiative and exploit opportunities guided by the commander's intent. A common example of command-by-influence, or mission command, is the so-called "strategic corporal." The expression "strategic corporal" represents the idea that command decisions should be delegated to those closest to the action, even if they are junior in rank, because they have the best picture of how to achieve the commander's intent within their area of responsibility. Command-by-influence, or mission command, can only be effective, however, if subordinates have been adequately educated and trained to carry out their command responsibilities.[15] Czerwinski's contention that only command-by-influence systems are likely to be consistently successful in the twenty-first century is supported by a number of military communities, notably the US Marine Corps.[16]

THE PIGEAU-MCCANN COMMAND FRAMEWORK Pigeau and McCann devised their framework to address the lacunae in theoretical study of command in a military context and have begun to evaluate their framework using data gathered from Canadian military operations.[17] They note that whether involved in disaster relief, peacekeeping operations, or war, the CF deal in human adversity. Inevitably, the CF respond to and resolve this adversity through human intervention. Any new theory of c^2 must therefore assert the fundamental importance of the human as its central philosophical tenet. It is the human – for example, the CF member – who must assess the situation, devise new solutions, make decisions, coordinate resources, and effect change. It is the human who must initiate, revise, and terminate action. It is the human who must (ultimately) accept responsibility for mission success or failure. All c^2 systems, from sensors and weapons to organizational structures and chain of command, must exist to support human potential for accomplishing the mission. For example, c^2 organizations that are intended to allocate authorities and define areas of responsibility should facili-

Table 2.1 Command and control as actions

Commanding	Controlling
To create new structures and processes (when necessary)	To monitor structures and processes (once initiated)
To initiate and terminate control	To carry out pre-established procedures
To modify control structures and processes when the situation demands it	To adjust procedures according to pre-established plans

SOURCE: Pigeau and McCann, "Re-conceptualizing Command and Control," 56.

tate the coordination of human effort to achieve mission objectives. If the organization hinders this goal – for example, by confusing lines of authority or by imposing excessive bureaucracy – then the human potential necessary for accomplishing the mission is also compromised. The challenge, then, becomes one of specifying those aspects of human potential that should guide c^2 development.

Pigeau and McCann first distinguish the concept of command from that of control, giving pre-eminence to command. They then link the two concepts together in a new definition of c^2.[18]

They define key terms as follows. Command is "the creative expression of human will necessary to accomplish the mission," and control is "those structures and processes devised by command to enable it and to manage risk. The function of control is to enable the creative expression of will and to manage the mission problem in order to minimize the risk of not achieving a satisfactory solution. The function of command is to invent novel solutions to mission problems, to provide conditions for starting, changing and terminating control, and to be the source of diligent purposefulness."[19] The functions of command versus control are shown in Table 2.1.

Pigeau and McCann's definition of command, which is markedly different from the standard NATO definition, is *the creative expression of human will necessary to accomplish the mission*. Without creativity, c^2 organizations are doomed to applying old solutions to new problems, and military problems are never the same. Furthermore, without human will there is no motivation to find and implement new solutions. For example, rarely does the slavish adherence to rules and procedures, devoid of creativity, produce effective organizations. Indeed, as will be seen elsewhere here, navies have traditionally avoided "doctrine," fearing it would restrict the initiative of their captains at sea. And as most labour unions know, a good method for hampering organizational effectiveness is to "work to rule," or to follow only "the letter of the law." Command, therefore, needs a climate of prudent risk taking, one where individuals are allowed to tap inherent values, beliefs, and motivations to marshal their considerable creative talents towards achieving common goals.

It follows from their definition that all humans have the potential to command; put another way, command is an inherently human activity that anyone, if they choose, can express. To limit command only to those individuals who have been bestowed with the title of "Commander" begs the question of what command is in the first place. Notice that this definition allows even junior non-commissioned members (NCMs) to command. If, through their will, they are creative in solving a problem that furthers the achievement of the mission, then they have satisfied the requirements for command.

But if all humans can command, on what basis do Pigeau and McCann differentiate command capability? What differentiates the private from the general officer? What key factors influence the expression of command? To address these questions, Pigeau and McCann have further refined the notion of command in proposing the concept of "effective command," which they define as "the creative and purposeful exercise of legitimate authority to accomplish the mission legally, professionally and ethically."[20] This definition highlights the notion of legitimate authority as the basis of effective command in the military. Even though all humans can command, according to their definition of command, the exercise of command by those not in positions of legitimate authority would probably not be deemed effective command in a military context. In this volume, the term "command" is used to denote "effective command" using the Pigeau-McCann definition.

DIMENSIONS OF COMMAND To elaborate further on their concept of command, Pigeau and McCann propose that command capability can be described in terms of three independent dimensions: competency, authority, and responsibility.

Command requires certain competencies so that missions can be accomplished successfully. For most militaries, *physical* competency is the most fundamental, one that is mandatory for any operational task, from conducting a ground reconnaissance to flying an aircraft. The second skill set, *intellectual* competency, is critical for planning missions, monitoring situations, reasoning, making inferences, visualizing the problem space, assessing risks, and making judgments. Missions, especially peace support missions, can be ill-defined, operationally uncertain, and involve high risk. Command under these conditions requires significant *emotional* competency, which is strongly associated with resilience, hardiness, and the ability to cope under stress. Command demands a degree of emotional "toughness" to accept the potentially dire consequences of operational decisions. Finally, *interpersonal* competency is essential for interacting effectively with one's subordinates, peers, superiors, the media, and other government organizations. These four aspects describe the broad set of competencies necessary for command.

Authority, the second dimension of command, refers to command's domain of influence. It is the degree to which a commander is empowered to act, the scope of this power, and the resources available for enacting his or her will. Pigeau and McCann distinguish between the command authority assigned from external sources and that which an individual earns by virtue of personal credibility – that is, between legal authority and personal authority. Legal authority is the power to act as assigned by a formal agency outside the military, typically a government. It explicitly gives commanders resources and personnel for accomplishing the mission. The legal authority assigned to a nation's military goes well beyond that of any other private or government organization; it includes the use of controlled violence. Personal authority, on the other hand, is that authority given informally to an individual by peers and subordinates. Unlike legal authority, which is made explicit through legal documentation, personal authority is held tacitly. It is earned over time through reputation, experience, strength of character, and personal example. Personal authority cannot be formally designated, and it cannot be enshrined in rules and regulations. It emerges when an individual possesses the combination of competencies that yields leadership behaviour.

The third dimension of command is responsibility. This dimension addresses the degree to which an individual accepts the legal and moral liability commensurate with command. As with authority, there are two components to responsibility, one externally imposed and the other internally generated. The first, called extrinsic responsibility, involves the obligation to public accountability. When a military commander is given legal authority, there is a formal expectation by superiors that he or she can be held accountable for resources assigned. Extrinsic responsibility taps a person's willingness to be held accountable for resources. Intrinsic responsibility is the degree of self-generated obligation that one feels towards the military mission. It is a function of the resolve and motivation that an individual brings to a problem – the amount of ownership taken and the amount of commitment expressed. Intrinsic responsibility is associated with the concepts of honour, loyalty, and duty, those timeless qualities linked to military ethos. Of all the components in the dimensions of command, intrinsic responsibility is the most fundamental. Without it, very little would be accomplished.

Pigeau and McCann's human-centred definition of command is a powerful tool for deducing some organizational principles. However, the careful reader will notice that simply specifying command characteristics is insufficient for completely describing c^2. How can one facilitate and support, for example, command expression? Under what conditions does the creative expression of will best manifest itself? Alternatively, unbridled creativity can lead to uncoordinated activity and organizational chaos. Under what conditions should the creative expression of will be limited or channelled? The answer to these questions is control. Command must execute to support and facilitate creative command, while controlling command creativity. Indeed, much of organizational theory can be seen as the attempt to establish the optimum balance between these two extremes.

As we have seen, Pigeau and McCann defined control as *those structures and processes devised by command both to support it and to manage risk*. Structures are frameworks of interrelated concepts that classify and relate things. The military environment encompasses a host of control structures (e.g., chain of command, order of battle, databases for describing terrain, weapon systems, organizations, etc.). Structures are attempts to bound the problem space and give a context within which creative command can express itself. For example, an organization's mission statement is a strategic-level structure whose purpose is to give long-term guidance to all members (including managers) in how to apply and channel their motivation and creativity. Once stable structures have been established, processes can be developed to increase efficiency. Control processes are sets of regulated procedures that allow control structures to perform work. They are the means for invoking action. Military rules of engagement (ROE), for example, are formal processes for regulating the use of power – for specifying the way in which military structures (e.g., soldiers, battle groups, and squadrons) are permitted to achieve their objective. Process increases speed of response and reduces uncertainty.

Knowing which structures and processes to invoke to achieve operational success is a key issue for command. Recall that Pigeau and McCann's definition specifies that control is *devised by* command. Structures and processes come into existence only through some creative act of human will. What are the guidelines for knowing when new control systems should be developed, or when existing control systems should be allowed to continue? Their definition specifies two broad guidelines. First, structures and processes should exist to support command. They should facilitate (or at least not hinder) the potential for creative acts of will. They should facilitate (or at least not hinder) the expression of competencies (physical, intellectual, emotional, and interpersonal). They should clarify pathways for legal authority; they should encourage (not impede) the opportunity to establish personal authority. And finally, they should encourage the willing acceptance of responsibility while at the same time increasing motivation in military members.

The second criterion for knowing when control should be invoked is whether it promotes the management of risk. Pigeau and McCann define risk loosely as anything that jeopardizes the attainment of the mission. This includes uncertainties related to personnel (including the adversary), uncertainties in the environment (e.g., weather, terrain, etc.), and the unbridled expression of creativity, since such expression may lead to chaos. Imposing an elaborate control structure and process is one way to reduce risk; however, this would come at the expense of inhibiting command creativity – creativity that inevitably is needed for solving new problems.

A tension exists, therefore, between the two reasons for creating control: to facilitate creative command and to control command creativity. Getting the balance right is a perennial challenge for most organizations. Pigeau and McCann suggest that as a general strategy, militaries should give priority to facilitating creative command. Mechanisms for controlling command creativity should then be used wisely and with restraint.

Their definitions of command and of control (as separate concepts) were designed to highlight a military's most important asset:

the human. However, a military is not simply a collection of independent individuals, each of whom pursues his or her own interpretation of the mission. Militaries are organizations for coordinated action, for achieving success by channelling the creative energies of their members towards key objectives. It is this important feature of military capability that Pigeau and McCann emphasize in their new definition of c^2: c^2 *is the establishment of common intent to achieve coordinated action.* Without coordinated action military power is compromised. Without common intent coordinated action may never be achieved. In their work Pigeau and McCann have specified some of the issues that must be addressed to elucidate common intent. They include a definition of intent itself (i.e., aim or purpose with associated connotations), an identification of two types of intent (explicit and implicit), and the mechanisms for sharing intent among military members, particularly between superiors and subordinates.

The key concept in their definition is *intent*, that is the set of connotations associated with a specific aim or purpose. When a commander gives the order to "Take Hill X by 1300 hours," he means not only take Hill X explicitly but also: "Take Hill X while making effective use of your resources, without killing innocent civilians, etc." Thus the commander's intent is made up of two components. The first is *explicit intent*, that part that has been made publicly available through orders, briefings, questions, and backbriefs. It includes communications that can be written, verbalized, or explicitly transmitted. But it is impossible to be explicit about every minute aspect of an operation. For expediency's sake, some things (actually most things) are left uncommunicated. Thus explicit intent carries a vast network of connotations and expectations – the *implicit intent*. Implicit intent derives from personal expectations, experience resulting from military training, tradition and ethos, and deep cultural values. It may be impossible to vocalize much of implicit intent. And it is usually acquired slowly, through cultural immersion or years of experience. Finally, common intent consists of the explicit intent that is shared between a commander and subordinates immediately before or during an operation, plus the (much larger) operationally relevant, shared implicit intent that has been developed over the months, and even years, before the operation.

as their subordinates (e.g., pilots) or by leaders who have a significant specialized knowledge of the jobs their subordinates perform (e.g., the seamanship skills of the naval officer). This type of leadership is critical in the navy and air force, where in every second at sea or in the air those on board ships and aircraft depend on technology (and by extension the technical ability of the crews and their leaders) for their very survival, not just their ability to fight.[5]

At the same time, there remains an important human dimension to naval leadership. Unlike aircrew, naval officers must live and work in close confines with their ratings, and especially on long sea voyages they find themselves in a leadership position day in and day out for months on end. The underlying balance is captured nicely in a phrase from the Canadian Navy's strategic vision document, *Leadmark*: "Warships, in all of their technological and sociological complexity, have been upheld throughout history as evidence of the achievements of a particular age."[6] That could explain why navies have usually been at the forefront of technical change, even though the stereotype of naval culture is of a service that is resistant to change.

The inherent tensions of technology versus the human dimension, as well as the characteristics of naval command described above, are very much evident in the philosophy of NCW. Recall that Arthur Cebrowski was a vice-admiral serving as president of the US Naval War College at the time he laid out the concept of NCW, and also that it was from this position that he was subsequently selected by US Secretary of Defense Donald Rumsfeld to head the Office of Force Transformation.

NAVAL STRATEGIC THEORY: A GLOBAL CONCEPT OF OPERATIONS
Naval warfare has not attracted the same level of analysis as has land or even air warfare.[7] The best-known proponent of sea power, Captain Alfred Thayer Mahan, initially codified his principles only as an afterthought on the insistence of his publisher, who hoped that *The Influence of Sea Power on History, 1660–1783*, which was published in 1890 as a primarily historical work documenting British success as a maritime power in dominating its continental rivals, might benefit from contemporary public controversy over future naval construction.[8] Appearing in what was very much an intellectual vacuum, his ideas came to dominate naval thinking for the next 100 years. Indeed, it is instructive that the major study *Makers of*

Modern Strategy includes Mahan as the only theorist of sea power.[9] His approach was essentially Jominian: he determined that maritime power derives from the dominating geographical position and economic and demographic conditions of the host nation, and he advocated that control of the seas could be gained most decisively through battle, rather than "secondary" strategies such as commerce-raiding or keeping a fleet in being. These notions held up quite well in an era of great power rivalry, as the French analyst Philippe Masson notes: "[despite] his not according sufficient importance to combined operations, and the confusion caused in the minds of certain military thinkers by the First World War, which was marked by the absence of decisive battle, Mahan's thinking was generally confirmed by the 1939–1945 conflict."[10]

Initial thinking after the Second World War was that nuclear warfare rendered such traditional concepts invalid; later, as the Cold War took hold, it seemed to point instead to "the importance of an area which Mahan had not developed at all: the political uses of sea power in peacetime."[11] By the 1970s, however, an appreciation that conventional weapons were not irrelevant to the nuclear age, as well as the emergence of an ocean-going Soviet surface fleet, led American naval leaders to conclude that "the service should strive for superiority at sea against the Soviets and ... think in terms of forward, offensive operations as the most effective means to employ the navy to achieve the nation's broad defense policies."[12] This renaissance in naval thinking found its intellectual foundation in Mahan, but where Mahan envisioned opposing naval forces that might engage in battle at various key points around the world, the US Navy's "Maritime Strategy" was truly global in perspective. In testimony before the US Congressional House Committee on the Armed Services in 1979, Chief of Naval Operations Admiral Thomas B. Hayward described his strategic approach:

> ... a NATO-Warsaw Pact conflict would "inevitably be worldwide in scope." To meet the challenge, the Navy would have to be "offensively capable ... The geographic range of the Navy's responsibilities is too broad, and its forces far too small, to adopt a defensive, reactive posture in a worldwide conflict with the Soviet Union." The U.S. Navy would have to carry "the war to the enemy's naval forces with the objective of achieving the earliest possible destruction of his capability to interfere with

our use of the sea areas essential for support of our own forces and allies." Carrier battle groups, representing the American technological lead over the Soviets, were "optimally suited for the execution of this strategy."[13]

Although the Maritime Strategy was never put into practice, by pressuring the Soviets to engage in a prohibitively expensive naval arms race, it was a major contributor to the economic collapse of the Soviet Union. With Western command of the seas assured, the end of the Cold War seemed once again to prove the validity of Mahan's thinking. The dawn of a new age of geostrategic uncertainty, however, ironically led to a rediscovery of Mahan's British contemporary, Sir Julian Corbett, also a naval historian-turned-strategist. More Clausewitzian in his approach, Corbett acknowledged the central role of the decisive battle, but he also saw "other important aspects of the use of sea power in the need to maintain seaborne trade and in the ability to threaten enemy coasts."[14] This thinking seems almost tailor-made for the present age,[15] and it has become the foundation for present-day Western naval strategic thought, such as it exists.

That latter assessment is made not to disparage, but rather to recognize (as noted at the beginning of this chapter) that navies generally have lagged behind the other services in codifying their strategic and doctrinal principles. Where the Gulf War of 1991 seemed to prove the continued validity of the Army's AirLand Battle concept and the exercise of strategic air power as epitomized by the US Air Force (these concepts are discussed in the chapters that follow), the US Navy was quick to appreciate that the tenets of its Maritime Strategy no longer applied and that a new theoretical construct for sea power was required to assure its strategic dominance and to make best use of it into the twenty-first century. The effort eventually produced *Forward ... From the Sea* (1994), which describes the unique capability of naval expeditionary forces to influence events in the littoral regions of the world. Allied navies soon followed this lead: the Royal Navy with a glossy update of its traditional *Fighting Instructions* as *The Fundamentals of British Maritime Doctrine* (BR 1806, second edition 1999), the Royal Australian Navy with a similarly styled *Australian Maritime Doctrine* (2000), and the Canadian Navy with *Leadmark: The Navy's Strategy for 2020* (2001).

Interestingly, other than certain aspects of the *Fundamentals of British Maritime Doctrine*, none of these documents describes naval warfare in anything resembling the detail of army doctrine, which reflects a general naval preference for sweeping guiding principles and a reluctance to waste effort on statements of the "obvious"; one is tempted to observe that these characteristics are also inherent in NCW, which is remarkably short on detailed theoretical foundation. However, other than technical procedural differences, all of the doctrinal statements listed above encompass a group of common war-fighting principles: in a complex age of undefined and asymmetrical enemies, maritime warfare would remain global in scope and contingent upon the bedrock of Western command of the sea; sea power is not geographically constrained like land power or of limited loiter endurance such as air power, nor dependent as are the other services upon fixed bases of operations; security of the sea lanes will allow the shift of focus of naval forces from the "blue water" of the open sea to the "green water" of the littorals, and that focus will aim to contribute materially to influencing the outcome of events in the decisive theatre ashore; greater autonomy of action is presumed for deployed commanders, contingent upon the information "situational awareness" to hand; and the increasing use of information technology and precision-guided munitions will enable the shift of concepts of warfare from ones based upon mass to those based upon manouevre.

The most recent – and forceful – exposition of this philosophy is the US Navy's *Sea Power 21*. In roundabout fashion, it has come to incorporate much of the language of NCW, as the opening phrases by Chief of Naval Operations Admiral Vern Clark demonstrate: "Sea-based operations use revolutionary *information superiority* and dispersed, networked force capabilities to deliver unprecedented offensive power, defensive assurance, and *operational independence* to Joint Force Commanders [emphasis added]."[16] The document concludes:

"Sea Power 21" is our vision to align, organize, integrate, and transform our Navy to meet the challenges that lie ahead. It requires us to continually and aggressively reach. It is global in scope, fully joint in execution, and dedicated to transformation. It reinforces and expands concepts being pursued by the other services – long-range strike; global intelligence, surveillance, and

duction of a multitude of new wide-area sensors (AWACS,* satellites, SOSUS,‡ etc.) that in turn threatened to overwhelm the OTC with more information than he and his staff could process. Simultaneously, the range of these new sensors in tandem with the reach of carrier aircraft equipped with guided munitions were expanding the scope of naval warfare to encompass literally oceanic breadths (i.e., an area of responsibility measured in hundreds of thousands of square miles). From all of this came the US Navy's new Maritime Strategy, which envisioned the management of a global naval war against the Soviet Union. One present-day paper, tracing the evolution of NCW, points to this period as the defining moment when the US Navy began to recognize the "dramatic doctrinal evolution occurring in the Army" in the development of AirLand battle and to begin its own shift from platform-centric mass to manouevre warfare, incorporating the predominant principles of initiative, speed, and long-range, precise firepower.[22]

Managing this level of complexity was a challenge for the best of admirals, even with the support of a full staff and the new specially designed command and control ships *Blue Ridge* and *Mount Whitney* (which entered service in 1970–71, assigned respectively to the Pacific and Atlantic commanders). The answer was the Composite Warfare Commander (CWC) concept.[23] This concept is recognizable today as entirely consistent with the Pigeau–McCann model in distributing "responsibility" and devolving appropriate "authority" downwards to "competent" subordinate commanders in each of the warfare areas (e.g., anti-submarine warfare [ASW], anti-surface warfare [ASUW], anti-air warfare [AAW], and strike), under the watchful eye of the OTC, who, having provided his overall "commander's intent," could use the range of modern communications at his disposal to monitor progress and exercise "command by negation" from his flagship. Although splitting the warfare responsibilities amongst subordinate commanders had begun to take place earlier over the years, the practice was inconsistent and not always clearly defined. In the early 1980s, the US Navy sought to institute a more formalized procedure with clear delegation of command

---

* Airborne Warning and Control Systems, such as the US Air Force's E-3 Sentry.
‡ Sound Underwater Surveillance System, a network of listening devices on the ocean bottoms.

and recommended it to NATO as an "experimental tactic."[24] By the end of the decade the CWC concept had been formally integrated as doctrine into NATO's primary Allied Tactical Publication, ATP-1. Through the years, various refinements were added, such as a series of operational tasking (OPTASK) formatted messages by which the OTC and his subordinate commanders could promulgate standing instructions and any modifications to them in a familiar and easy-to-read style.

With only slight modification, the CWC concept has been established as the basic organizing concept for command of coalition naval warfare. The *Multinational Maritime Operations Doctrine Manual* provides a general description:

> The grouping of units within a maritime Task Force will be based on the force's assigned mission and threat. Protecting the force while conducting sea control and power projection missions normally requires the selective delegation of warfare functions (i.e., ASW, AAW, ASUW, etc.) from the Officer in Tactical Command (OTC) to a subordinate commander. This process is embodied in the Composite Warfare Commander (CWC) concept.
>
> The CWC concept allows the Task Force Commander or OTC to delegate tactical command (TACOM) to a Composite Warfare Commander to conduct power projection operations, counter threats to the force, and conduct sea control operations. However, in practice, the OTC would normally retain the duties of CWC [see Figure 3.1]. Depending on the size and complexity of maritime operations, the CWC may subsequently delegate tactical control (TACON) of some or all of the warfare functions to subordinate commanders. When maritime operations involve more than one Task Force, each TF will retain its own CWC structure. To minimize mutual interference, the MNFC [Multinational Force Commander] will need to identify how TF warfare and sustainment activities will be coordinated within the maritime battlespace (e.g., establish a geographic area of operations for each TF or assign one TF to coordinate all TF activities). During a joint campaign, command of power projection operations such as strike and amphibious operations would normally be retained by the CWC of the carrier or amphibious TF. Functional responsibilities can also be assigned for selected operations such

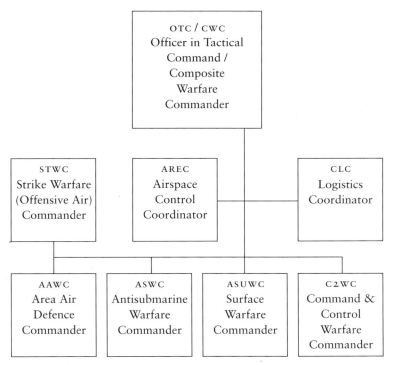

Figure 3.1 An example of Composite Warfare Commander (CWC) organization.

as maritime interdiction (MIO) or special operations (Spec Ops). A similar organization may be implemented for a multinational force. When directed by the MNFC, subordinate commanders will act as warfare or functional commanders within the multinational force.[25]

The CWC concept still retained the appearance of "command-by-direction" in its hierarchical task organization (see Figure 3.1); however, in breaking down the previous rigid centralized control and allowing for subordinate warfare commanders' intent, it conceptually shifted western C^2 practices to something more closely resembling command-by-influence.

THE CANADIAN NAVAL EXPERIENCE OF NETWORK ENABLED OPERATIONS This is an appropriate point at which to turn to the

Canadian naval experience with NCW. This presents a unique case study in that because of our close operational association with the US Navy, our Navy has been deeply involved in the practical implementation of the concept virtually from the beginning. However, as a junior alliance (and more recently coalition) partner to the US Navy, the Canadian experience differs in some fundamental ways. These in turn highlight points of vulnerability in the US Navy approach, while offering useful thoughts on the way ahead. Where a number of naval analysts foresaw a major problem with NCW being the ability of coalition partners to keep pace with the US Navy,[26] the practical experience of what might be called "the Canadian model" is that there is a place for small navies in NCW.

As a NATO ally, the Canadian Navy quickly became familiar with the CWC concept but regrettably was only able to employ it initially at a much lower level than intended, having lost its aircraft carrier in the late 1960s and consequently being consigned to the close-in ASW screen around the carrier (that normally being a function retained under the direct control of the OTC or the US Navy cruiser commanding the inner screen). However, the Canadian Navy did possess four modern DDH-280 Iroquois-class destroyers equipped with the latest computerized digital display command and control system, which was connected to the rest of the force by the revolutionary ship-to-ship-to-aircraft tactical datalink Link-11,[27] and, importantly, embarking two large Sea King ASW helicopters, which rationalized a 100-mile radius of operational control. The industrious Canadian staffs who found a new home in these "Sisters of the Space Age" (as the DDH-280s were styled when commissioned in the early 1970s) pressed to be delegated the subordinate ASW commander duties, and for their reward they quickly found themselves deluged with information, not unlike their senior American commanders, even within that area of specialization. A later commander of Maritime Command, Vice-Admiral Lynn Mason, recalls that as captain of *Iroquois* in 1981–82, the volume of operational message traffic led him to assign a senior ship's officer full-time to the task of filtering and digesting it into a manageable volume that he might absorb efficiently.[28] The effort was magnified at the task group level, requiring the embarkation in the modest Canadian destroyers of additional officers and senior ratings (and the supporting communications systems) to form a staff roughly equivalent to that of a US Navy battle group to manage the information flow. With dogged determination, however, by the late 1980s Canadian task group commanders were regularly exercising the ASW Commander function in major NATO exercises.[29] By the time of the Persian Gulf War of 1991, the level of staff expertise had received general recognition such that the commander of the Canadian task group was the only non-US Navy officer assigned a subordinate warfare commander role in that conflict, as commander of the Coalition Logistics Force.[30] That same commander discerned that the prevailing conditions of the ad hoc structure of coalition warfare were far different from the obligations of a formal alliance, which led him to suggest a redefinition of C^2 to mean "cooperation and coordination."[31]

In the meantime, other developments had pushed the Canadian Navy quite legitimately into the CWC role while also laying the groundwork for later participation in the network-enabled revolution. From the mid-1980s, several Canadian ships were equipped with towed array sonars, in anticipation of that being the primary underwater sensor of the new general-purpose Canadian Patrol Frigates then under construction. With passive detection ranges often in the order of hundreds of miles, these ships were most effectively stationed well outside the inner screen. Such over-the-horizon ranges, in contrast to active sonar ranges measured in thousands of yards, quite literally broadened the perspective of Canadian naval commanders. In another sense, they also were beyond the capacity of the line-of-sight UHF Link-11 datalink and longer-range HF frequencies that could follow the earth's curvature, because they proved unable to handle the flow-rates necessary for a reliable datalink (HF signals also were easily subject to enemy direction-finding and as such had to be avoided in most tactical situations). The obvious solution was acquisition of satellite communications (SATCOM) operating at UHF and higher frequencies (all line-of-sight into space and returned on a narrow undetectable "footprint"). Of the various NATO systems becoming available, Canada fatefully invested in the US Navy Fleet SATCOM because our towed array-equipped ships were working closely with US Navy forces in the prosecution of strategic ASW against Soviet ballistic missile-firing submarines. This new role was made possible, as discussed earlier, primarily through the unique high-level Canadian access to American communications and command procedures, and indeed Canadian ships commonly came to be equipped with standard fits initially of the JOTS (Joint Operational Tactical System) digital display and eventually its web-based successor, GCCS-M (Global Command and Control System, Maritime).[32] In return, the unique Canadian role in facilitating command of coalition naval forces in the Persian Gulf in 1990–91 validated the broader applications of this high-level access onto the international scene, beyond the strict defence of North America into the realm of coalition warfare. Where interaction with American naval forces had always been an accepted part of broader alliance duty for all navies within NATO, the Canadian Navy suddenly realized a much deeper meaning: what would soon become known as "interoperability with the US Navy" was appreciated to be a fundamental operating tenet, and within that, communications were a key element worth continued investment.

By the early 1990s, therefore, two of the essential building blocks for NCW were in place: the routine assignment of separate warfare commanders for AAW, ASW, and ASUW within the CWC concept had flattened traditional notions of command at sea, and the increasing use of Fleet SATCOM allowed real-time global communications at reliable data flow-rates. The Canadian Navy had enjoyed recent practical experience in both domains, and each in a fashion that solidified the unique relationship with the US Navy. Ironically, these two conditions came about at the same time that the geo-strategic situation driving American C^2 innovation changed fundamentally. In Canada, as within NATO more generally, there was an expectation that the collapse of the Soviet Union would lead to a "peace dividend" in the form of reduced military budgets. In the United States, however, the disappearance of a "peer competitor" did nothing to slow continued development of new information systems. The impetus now was coming from technology itself, as the computer revolution began to be felt at a more personal and hence immediate level in the early 1990s, with widespread access to the Internet and user-friendly web-based software functions.

The full implications of the computer revolution for command at sea were not truly appreciated anywhere in the early 1990s, let alone within the Canadian Navy, which was something of a trend-setter in the onboard use of personal computers in a variety of fashions. At a very basic level, the age of the Canadian ships available to deploy to the Persian Gulf had embarrassed the government into taking whatever steps were necessary to equip them properly, and the resulting free flow of money (albeit still at relatively modest amounts, given only three ships were involved) created a proliferation of personal computers acquired for use as word processors throughout the ships, which led in turn to a great leap in the computer literacy of Canadian sailors. More prosaically, the early 1990s saw the virtual rebuilding (one might call it the "transformation") of the Canadian fleet, with the introduction into service of the new-build Halifax-class Canadian Patrol Frigates (CPF) and the upgrading of the Iroquois-class destroyers through the TRUMP program (Tribal Update and Modernization Program). Since both programs were designed in the mid-1980s, computers were a major element in their equipment: for example, the basic platform machinery and combat systems of the CPFs were run by a revolutionary (for the times) distributed network of what were essentially a dozen Commodore 64s (i.e., 64 kilobytes times twelve, or less than one megabyte of total memory); TRUMP accomplished roughly the same for the destroyers. The command and control systems within both these classes, being of such recent design origin, were the envy of many other navies.[33]

As computers rapidly spread into wider use in the first half of the 1990s, similar developments were occurring in other navies, especially the US Navy. As access to the Internet also became generally available, one of the first to recognize the operational c^2 potential was the Pacific fleet commander based in Pearl Harbor, Hawaii, on whose associated staffs serve a number of Canadian naval officers. In particular, the commander task force 12 (CTF12), the Pacific theatre ASW commander, appreciated that a satellite-based classified military version of the Internet was the ideal medium through which he could solve a number of c^2 problems and coordinate the multiple inputs over the vast distances from the variety of units and assets under his command.[34] A collaborative effort with the defence industry supplier Orincon Corporation produced the "WeCAN" (Web-Centric Anti-Submarine Warfare Net) as a "real-time theatre and tactical level information sharing capability for USW (Under-Sea Warfare) collaborative planning and execution."[35] WeCAN first became operational in 1996, and from its beginning as a network for e-mail exchange and the sharing of common basic information posted to a centralized server, it immediately proved its worth. For example, it allowed new units joining an ASW operation to pull down pertinent information from the net (typically in the form of pre-formatted messages), rather than the ASW commander having to retransmit a large volume of messages; another function allowed units prosecuting a contact to post their acoustic and other sources of information for viewing and collaborative assessment by other units. Within the US Navy this eventually was imported to the developing SIPRNET (Secret Internet Protocol Router Network, a US Department of Defense-managed system to allow the sharing of classifed information among military personnel, with multiple levels of access) to allow access with the other branches of the US armed forces. Experiments were conducted with a lower-classified version known as COWAN (Coalition Wide Area Network) during Exercise RIMPAC 98, which involved various "Rim of the Pacific" nations in the summer of 1998, and this became the basis for a COWAN used by naval forces gathered in the Arabian Sea for Operation Enduring Freedom from the fall of 2001. With the availability of additional soft- and hardware technology such as chat rooms, WeCAN / COWAN has quickly grown into CENTRIXS (Combined Enterprise Regional Information Exchange System) and a collaborative command decision-making tool, the latest version of which is known as "c@s2" (Collaboration at Sea, Version 2).

Once again, the Canadian Navy was deeply involved with this process from its earliest days. In 1995, the Canadian government decided to deploy a warship to the Persian Gulf to assist in the enforcement of United Nations sanctions against Iraq. The ship designated was one of the new frigates, the Esquimalt-based HMCS *Calgary*, and was to be part of a US Navy Third Fleet carrier battle group (CVBG). In pre-deployment planning, it occurred to the various staffs that the modern weapons and communications systems of the frigate meant *Calgary* could effectively take the place of a US Navy ship, helping to relieve the manning pressures facing the American fleet as it downsized from the 600-ship Cold War navy. Such full "integration" into the battle group was contingent upon complete communications "connectivity," but even the high-level access enjoyed by the Canadian Navy was not sufficient to allow SIPRNET access. Recognizing the benefit to both parties, the new commander of Maritime Command, Vice-Admiral Lynn Mason, met to resolve the issue with the chief of naval operations, Admiral Mike Boorda (in Mason's words, "an old friend" – another indication of the continued importance of "human" networks).[36] Boorda agreed to Mason's proposal, and *Calgary* was fitted with all the required systems. The deployment proved to be a great success, becoming the model for a subsequent half-dozen similar deployments that took place before the fall of 2001 (roughly one a year, each for about a six-month period).[37]

An important aspect of these communications fits (which included not only the specialized displays and computer software but also the internal ship local area network [SHIPLAN] and external satellite components) was that they were obtained as "mission fits," that is, specific acquisitions for each particular ship for the designated mission. It is Canadian practice that mission fits of most equipment are removed on completion of a deployment, typically for transfer to a ship proceeding on the next mission. In this instance, the US Navy had developed sufficient incremental improvements to the SIPRNET / COWAN system by the time of each subsequent mission that it was better value to obtain a whole new system each time rather than expend the additional removal and installation costs; at the same time, however, the original basic fit remained sufficiently useful that it could be brought up to date through an end-of-fiscal-year "minor requirements" purchase.[38] In this fashion, by the summer of 2001, nearly half of the frigate fleet was fitted with network-enabling technology and, importantly, a large number of crews were familiar with its use. It had all been accomplished at significantly lower cost and in a much abbreviated period than with a formal capital acquisition program.

The fits and familiarity were not distributed evenly, however, because the majority of frigate integrations had been with Pacific Fleet battle groups, and only one ship from Canada's Atlantic fleet had occasion to experience the complete communications interoperability with the US Navy; this was *Charlottetown*, which was attached to a Second Fleet Surface Action Group (commanded from a cruiser and therefore significant also in operating at lower communications level than an aircraft carrier battle group). Off-setting that to a limited extent was the fact that the east coast destroyers *Iroquois* and *Athabaskan* had been fitted with a much larger scale of equipment for the half-year each had been employed as the flagship for Canadian Commodore David Morse's command of the NATO Standing Naval Force Atlantic (SNFL) in 1999–2000, although again this was not to the high connectivity level of the west coast. An assessment of that command tour offers the following overview:

The most recent operations have demonstrated that information management is becoming a complex affair. The USN [US Navy] is leading the Revolution in Military Affairs (RMA) at sea, and countries that wish to operate with the USN, in the littorals or elsewhere, must try to stay in step. This requires compatible high-speed and secure data links, use of the Internet for open source intelligence, internal ship and staff Local Area Networks (LANs), video teleconferencing between ships and formations, and other means of rapid and reliable communication. The Canadian Navy is well placed to conduct operations with the US and other Allied navies; however, a great deal of effort will be required to achieve Network-Centric Warfare (NCW), where all operational maritime units are "netted-in" to the operational picture. NCW will likely lead to centralized control of sensors and weapons to optimize capabilities, thus making the waging of maritime warfare more efficient. Ultimately, this enhanced "connectivity" will allow naval forces to "link in" more completely with both land and air forces that are also operating in the littorals.[39]

From this auspicious practical operational beginning, the involvement of the Canadian Navy soon came to include the theoretical development of NCW itself. Among those observing the increasing use of networked communications was Vice-Admiral Cebrowski, the US Navy's director for space, information warfare, and command and control (N6); in concert with a number of former colleagues elsewhere on the Pentagon's joint staff, he began to develop his vision of NCW, which was first published in January 1998.[40] He was able to exploit his anticipated shift to the Naval War College to include NCW as one of the experimentation elements for that summer's Global War Game (GWG), an annual modelling and simulation event that began in 1979 specifically to game the US Navy's Maritime Strategy.[41] Through the 1990s, the scenarios obviously had narrowed somewhat from global war with the Soviet Union, but attendance at the game had broadened to include the other close allies, Australia, Canada and the United Kingdom (AUSCANUKUS, commonly referred to collectively as "four-eyes"). Normally the game was staged in-house at the War College, but to add realism to the 1998 experiment Cebrowski included a real-time satellite link with the Third Fleet carrier battle group preparing to deploy to the

Persian Gulf – including the integrated frigate *Ottawa*. Such a level of activity, including Canadian and allied participation, continues to this day, leading to a fairly broad theoretical as well as practical familiarity with the concept.[42]

In the wake of the Al-Qaeda attacks on the United States on 11 September 2001, Canada deployed the east coast task group to support the global war on terrorism (widely known as Operation Enduring Freedom, the initial Canadian contribution was called Operation Apollo).[43] The Canadian Navy was the first major non-American force to arrive in-theatre, and it quickly found itself charged with the significant undertaking of exercising command over other coalition naval forces as they arrived in the Arabian Sea theatre of operations (four Canadian commodores served in succession over the eighteen-month period from November 2001 through June 2003). The appointment as warfare commander for the Arabian Sea theatre of operations, CTF 151 (commander task force 151) was arguably the first true exercise of operational-level command by a senior Canadian officer since the Second World War. It could not have been accomplished as successfully or as professionally as it was but for the employment of networked operations. A similar coalition task force, CTF 150, operated in the Horn of Africa region and was commanded in rotation by continental European Union allies, but to nowhere near the same level of connectivity with the US Navy or consequent operational success.

Just as the Arabian Sea was by now a familiar theatre of operations for the Canadian Navy, the progressive engagement in network-enabled operations over the previous decade facilitated the maintenance of completely integrated communications connectivity with US Navy commanders, specifically commander Fifth Fleet (NAVCENT) in Bahrain and the others embarked at-sea in the carriers. To complete the c^2 links, Canadian commanders at national headquarters in Ottawa and the command detachment at US Central Command (CENTCOM) headquarters in Tampa, Florida, were outfitted with necessary equipment and channels, as was the detachment commander of the Aurora long-range maritime patrol aircraft at Camp Mirage (the Canadian air base at an undisclosed location on the Arabian peninsula). For those air force and army officers not yet familiar with this level of networked operations, it was eye-opening experience. The full range of network-enabled capabilities comprised a dizzying array of acronyms and systems (at

least to the lay reader), in themselves testament to the complexity of
the range of information at the disposal of the modern naval commander: the "secret" level COWAN, with its cross-linked web pages,
e-mail, and "Sametime Chat" features; MCOIN III (the most recent
web-based version of the Maritime Command Operational Information Network, a classified national wide-area network similar to
the American SIPRNET, with COWAN and now CENTRIXS residing on
it); and the Link-16 and Link-11 tactical datalinks, as well as GCCS
to maintain the "recognized maritime picture."

The global ranges involved, and the enormous bandwidth "pipe-
line" needed to maintain these networks, could only be assured by
ready access to several SATCOM channels, which therefore also
became the limiting factor. To extend the connectivity throughout
the task force, the Canadian task group commanders pushed other
coalition members to adopt COWAN. The US Navy was quite happy
to allow firewalled access to lower levels of their classified networks, but even longstanding NATO members ended up declining to
engage, mostly because of the costs of obtaining the required technology and renting the satellite channels. Those few that did join
COWAN (such as the British and Australians) chose to attempt to do
so on satellite channels rented for only short time blocks each day,
but ultimately they found that partial engagement operationally
limiting, forcing them into discrete roles: the Royal Navy commander chose to maintain his task group as a separate entity, which
allowed direct support to the US Navy but reinforced Britain's separation from the rest of the so-called "Coalition of the Willing."[44]
When the US Navy delegated command of interception operations
in the northern Persian Gulf area to the Royal Australian Navy, the
Australian task group commander found he had to embark in the
US Navy Aegis destroyer assigned to his group to gain access to the
communications and displays needed to perform the role.[45] The
Canadian commanders were not so constrained, but they did appreciate the limitations of the forces under their command. However,
the coalition forces also constituted the bulk of the forces assigned
to the area, and to ensure their most effective operational employment, the Canadians pressed hard for their access to the vital information exchanges. Eventually a modest short-range network was
established for the non-SATCOM-fitted coalition members through
the mediums of HF Battle Force E-mail (BFEM) and Link-11 tactical
datalink (TADIL). But as Commodore Lerhe observed, "We never

got BFEM to carry longer than 30–40 nautical miles, so calling it an 'Arabian sea network' is a bit of a stretch."[46]

Although the maintenance of these networks was contingent upon the capability to perform the range of warfare activities, which defined CTF 151's mission as a classic exercise of sea control, in fact the networks had little to do with the warfare exercise of power projection envisioned by the proponents of NCW. This distinction was underscored by the very purpose of CTF 151, which was to facilitate the continued engagement in the war against terrorism of coalition members who disagreed with the American prosecution of the war against Iraq. As such, there was a clear separation of activities between the overt war fighting of Operation Iraqi Freedom and the picture compilation and maritime interdiction of the ongoing Operation Enduring Freedom. That these activities "lower" down the military scale were just as contingent upon the skilled use of network-enabled capabilities can be illustrated by two anecdotes from opposite ends of the NEOps spectrum. These anecdotes serve also to underscore that the object of NEOps should be to "wire" not every sailor but only those with a role that can specifically benefit from such a capability; neither does NEOps have to lead to the micromanagement of field operations by higher authority.

First, at the highest level, was the availability in the form of real-time chat of a network by which senior commanders could discuss the nuances of an unfolding operation. At one point shortly after the invasion of Iraq began, a Canadian ship intercepted an Iraqi vessel in the southern Persian Gulf thought to be engaged in mine laying. Although this was a clear violation of international law, the task force commander knew the sensitivity of the Canadian government to any activity involving Iraq. The details of the operation are unnecessary to this narrative (the Iraqi vessel was boarded and found to be not so involved), but the method of coordinating it is intriguing: to ensure that senior commanders of all the appropriate authorities were engaged to be made aware of events and so that their timely advice could be sought, the on-scene Canadian task force commander established a chat net including himself, the US Navy commander of NAVCENT in Bahrain, the Canadian general in CENTCOM Tampa, and the deputy chief of the Defence Staff (DCDS) in National Defence Headquarters (NDHQ) in Ottawa.[47]

The other example, at the lower extreme, points to the potential for network-enabled capabilities to facilitate the actual physical conduct of an operation. Coalition ships were often cued by intelligence sources as to which vessels should be stopped and searched for suspected Al-Qaeda and Taliban members, but they still faced the enormous task of identifying who out of those on board (often numbering in the dozens) might be the individual sought. Ordinarily, such a process could take hours, to the immense discomfort (in the 40 degree Celsius heat and 100 per cent humidity of the Arabian Sea) of all involved. The Canadian crews soon perfected a technique whereby a member of the boarding party equipped with a digital camera would take photos of the possible Al-Qaeda suspects, to be e-mailed to NAVCENT for analysis against their continuously updated databank, and then direction was provided back to the ship via chat, often within minutes. Not only did this save the ships taking large numbers of suspects for detention in the cramped onboard spaces (only to have the majority of them released), but it saved the ships from having to maintain an onboard databank, the constant revision of which was enough to occupy a large part of the NAVCENT staff. The longest part of this whole process as practised in 2002–03 was the time taken to transfer the boarding officer's camera back to the mother ship; this has since been improved by the next generational leap in technology, by equipping the boarding party with a handheld tablet on which the digital photos are downloaded and sent to the mother ship by a wireless router.

The Canadian task group commanders saw the employment of NCW principles in the instantaneous worldwide communications and situational awareness to allow tactical actions based on blended inputs. As one of them, Commodore Eric Lerhe, put it, "The Task Group Commander embarked in a Canadian destroyer enjoyed a level of C3I unmatched outside of a US Navy cruiser."[48] One might observe that they put it to even better effect by working hard also at innovative and adaptive uses of other technology, such as expanding the technology net to include those outside of it. Indeed, the Canadian Navy has recognized its "force multiplier" potential by acting in a "Gateway C4ISR" capacity between the US Navy and less well-equipped coalition members.[49] The challenge remains the ability to maintain the pace being set by the US Navy. As a senior Canadian naval officer observed, "Technological solutions are being developed to overcome these obstacles, however a restrictive information-sharing culture in the US is proving to be as difficult as the technical one. Until these problems are resolved, the Canadian

Navy's necessary vision of seamless technological procedural interoperability with the US Navy will remain highly problematic."[50]

Indications of mixed progress came in the course of RIMPAC 2004, which saw the Canadian Pacific Fleet commander, Commodore Roger Girouard, in charge of the coalition force. That was a familiar role (he had also served as CTF 151 in the Arabian Sea), and he managed it from his flagship *Algonquin*, which had been greatly improved over even just the year since Operation Apollo with twenty-eight CENTRIXS / COWAN "drops" (the term given each computer terminal in the onboard local area network); in contrast, the entire USS *John C. Stennis* battle group, also in the exercise, had only twelve drops.[51] Overall, the exercise was generally a success on the level of information management, with each participant linked into CENTRIXS. Still, because of the restrictive information-sharing culture, which was complicated by the separate bilateral relationships of the US Navy with the participating regional partners, the exercise commanders knowingly violated one of the cardinal principles of networked operations, in using unique channels with each separate exercise participant: for example, CENTRIXS–4 Eyes (USN/CDN/AUS/RN), CENTRIXS-J (USN/Japan), CENTRIXS-C (USN/Chile), CENTRIXS-R (USN/Republic of Korea), and so on.

Perhaps this will remain a necessary consequence of ad hoc "coalitions of the willing" for the foreseeable future. The redefinition of C^2 in coalition operations to mean "cooperation and coordination" reflects the reality of command in the future where coalition operations may predominate. The new cooperation and coordination paradigm appears to emphasize leadership or influence behaviours among peers over traditional concepts of command involving the exercise of authority over subordinates. Therefore, in coalition operations the leadership concepts of emergent leadership and distributed leadership may be more useful than concepts of authority. In fact, one might see the high reputation that senior Canadian naval officers have earned in certain operational command positions as a type of emergent leadership based on three subclasses of personal power (i.e., expert, referent, and connection), rather than position power.[52]

NETWORKS AND NAVIES An important issue that has arisen in the application of networks to navies is the effect of networks on naval command accountability. One of the best recent analyses on this topic was written by retired US Navy Captain Chris Johnson, formerly the director of the prospective commanding officer/prospective executive officer course at the US Navy's Surface Warfare Officer School. His key ideas are summarized to put NCW in a naval context.

Johnson starts his analysis by reminding us that in the navies of previous centuries commanding officers at sea were charged with making decisions independently because there was little real-time direction from senior officers ashore. He states: "this empowerment to act alone was a unique feature of naval service, a point of pride that distinguished us as different – perhaps even more capable and responsible – than our sister services." Yet with this independence and trust came "an all-encompassing code of accountability" for how and why command authority was used.[53] Quoting former US Navy Chief of Naval Operations Admiral Vern Clark, Johnson notes that "'Accountability does not equal figuring out who to punish when something bad happens. It does mean holding our people – and particularly our commanders and those with the most responsibility – to account for their actions.'"[54] "The core of accountability," according to Johnson, is the premise that because they are given "the authority and the resources to achieve their missions and preserve the safety of their commands," commanders can be held accountable. However, as communications have improved, especially from the last decade of the twentieth century onwards, the ability of commanding officers to control their "command's destiny" has gradually and almost imperceptibly been eroded.[55]

Now "an almost unfathomable cast of people," whom Johnson calls "peripheral actors," can influence the ability of commanding officers at sea to achieve their missions. NCW has caused "control of a ship's destiny" to migrate in two directions: to the periphery of the network, and up the chain of command. These effects of NCW are a result of deliberate US Navy policies, and are, therefore, the way of the future: "The issue, then, is not whether net-centricity is good but how the Navy will hold people on the periphery accountable, side by side with the commanding officer and his crew, for their impact on the success or failure of the ship."[56]

The implementation of "net-centric accountability" is vital if the US Navy is to remain an organization "founded on principles of justice and merit." Therefore, the navy must develop ways to track accountability throughout the network to maintain the confidence of its officer corps in the fairness of the system, Johnson argues. Since the US Navy's command system is based on the principle that accountability is balanced with authority, if authority is being dispersed by NCW, then so must accountability. Tracking accountability also has the purpose of "institutionalizing self-learning" in the new NCW navy, for a learning organization must be able to receive accurate inputs for effective learning to occur.

Johnson also addresses the migration of control up the chain of command through new technologies that permit instantaneous communications between commanders ashore and ships at sea. While he acknowledges that, for good reason, in the past the US Navy has been unwilling to hold senior officers publicly accountable for every action that happens under their command, if through NCW they become directly involved in command decisions, then they must also share some of the accountability.[57]

Four suggestions are offered by Johnson for the US Navy to adapt its command system to the NCW environment: 1) acknowledge that the nature and breadth of accountability has fundamentally changed and hold all persons and even systems, not just senior officers, accountable for their actions; 2) resolve that accountability must be tracked to every node in the system; 3) develop ways to track accountability so that the "where, when and who associated with key decisions" is known; and 4) develop a new accountability paradigm that is relevant to the NCW environment.[58]

This discussion of technology and command points to the great debate as to whether concepts like NCW will work effectively or not. To a large extent that will depend upon organizational culture as much as technology. However, because NCW originated as a concept within the US Navy, and its development has been followed closely by the Canadian Navy, the issue is especially important to this study.

RECOGNIZING THE HUMAN DIMENSION The Canadian naval networked operations model described earlier does not fully answer the accountability concerns raised by Johnson but, interestingly, it goes some way towards doing so by recognizing the importance of the human dimension as a means of rationalizing the tendency towards excessive reliance upon technology. There are several elements to this.

Fundamentally, human-centred networks are the basis of the Canadian naval command style, which is seen primarily in the predisposition to engage the widest variety of coalition members in task force composition and then to ensure their effective participation in any operations. In facilitating those operations, Canadian naval commanders appreciate that information technology is an enabler, not an end in itself, which leads them to choose the technology appropriate to the task at hand; in the case of command and control, this has come to incorporate a broad range of network-enabling capabilities. Indeed, a recent naval command study determined that "Canada's national culture with its traditions of bilingualism and multiculturalism; Canada's military culture with its history of alliance and UN operations; and Canadian naval culture based on operational and command competence, enlightened leadership and management techniques, *and a judicious exploitation of available technology*, make the Canadian Navy's command style a model for coalition operations [emphasis added]."[59]

The initial implementation of web-based information technology at sea has been a process of generally incremental and ad hoc modifications to existing systems.[60] This slowly building familiarity has allowed for a fair degree of experimentation, and indeed necessitated it, since the core level of experience – ordinarily the basis for the development of naval knowledge – was in this instance nonexistent. To be sure, not all the possibilities have been explored, but after nearly a decade of in-service employment, it has become evident to users that there may be difficulties in achieving the objectives of NCW in their purest form. For example, as discussed in more detail below, there are many interpretations as to just what is meant by the self-organizing NCW principle of "synchronization." While the related concept of self-synchronization is often used to refer to interpersonal activities and coordinated actions in terms of group activities and dynamics, there is reason to believe that self-synchronization may not work in the same way within complex systems. Indeed, initial experimentation in the naval environment supports the notion that there is a physical limit to the capacity of the normal human mind to sort, filter, and digest large quantities of information. As the critical element in the complex naval command and control system, human beings require a fundamental organizing principle. In the networks now developing in service, this is being refined in a number of practical and workable ways. To begin, there is a growing attempt to confine the range of distinct nets to a limited number, and work is being done to blend as many as practicable into single outputs or displays; a prime example is the Canadian-developed Automated Data Systems Integrator (ADSI), which does just what its name implies, with datalink computer graphics shown over top of the ship's processed radar and other source video. Within web-based systems such as MCOIN, COWAN and CENTRIXS, there is a determination to define an identifiable and logical organization, for example, by grouping information using the standard NATO formatted messages or by restricting chat nets to "Buddy Lists" based on the CWC organization in effect for an exercise or operation. The Canadian Navy has taken the additional step of promulgating a TACNOTE (tactical procedure note) codifying various aspects of the "Collaboration at Sea" (C@S) system. Carefully adding the caveat, "This is not approved doctrine," the purpose of the TACNOTE is described as follows:

> It is recognized that the continued pace of change in C4I equipment and software in use will result in dramatic and continued changes in the exchange of information at sea, also known as Collaboration at Sea, or C@S. This TACNOTE has been structured to provide current information on existing systems to new users and to serve, through the TACNOTE AMEND process, as a repository for information on new systems, equipment fits and lessons learned as they are brought online ...
>
> In the past year, COWAN-A has moved from a connectivity trial of limited ability during RIMPAC to a full-fledged warfighting network in the Persian Gulf ... The ability to effectively exchange information via networks is still young and its full potential has not yet been fathomed ... Technology is allowing us to share more information at a much higher volume than ever before, but this can quickly overwhelm the end user due to its sheer volume. Information management practices and technologies must continue to expand and develop in order to keep up with the sheer volume of information available to the warfighter.[61]

Perhaps it can be said that the practitioners of network-enabled operations have indeed self-synchronized, but into a recognizable order.

A very recent development in the field of information management practices is recognition that ultimately a human is the best
organizer of information required by other humans. Just as Vice-
Admiral Mason in the early 1980s appointed a senior ship's officer
to sort and filter the volume of message traffic into something he
could digest, the Canadian Navy has established the specialist classification of information management director (IMD) as one of the
warfare subspecialties for mid-level officers and senior NCMs of the
combat trades in the Canadian Navy (the other ones are anti-submarine warfare, above-water warfare, and navigation; the previous
communications subspecialty has been included in the IMD). All
duties and responsibilities of the IMD revolve around three primary
functions, as described in these excerpts from the terms of reference
for the position:

- *Data collection.* As a data miner, the IMD searches and retrieves
  information from the increasingly large quantity of sources, both
  aboard and ashore, while on watch in deployed vessels. These
  sources may be in printed publications held aboard, files held
  within indigenous networks aboard ship or the Task Group, or
  in networks ashore. The information will typically reside in a
  myriad of sources, classified, unclassified, commercial, military,
  academic, and government – sometimes being pulled (e.g., web
  servers) and sometimes requested (e.g., Request for Information
  counters). The IMD determines (based on connectivity confidence,
  file size, perishability and other considerations) whether the information is to be copied into shipboard servers or bookmarked /
  linked.
- *Data rationalization.* Given that the sources are so broad and
  diverse, the IMD must be able to convert the material into useable
  formats. Typically, this will mean conversion to ensure Microsoft
  Office suite compatibility, but can also include other considerations, such as material classification. IMDs will be trained to
  work with a larger scope of products than most other operators,
  as they will be required to standardize the material to prepare it
  for internal use or external dissemination.
- *Data dissemination.* The IMD will be the primary point of content
  for material being uplinked from the Ops Room, such as "end-of-
  action" reports, but more importantly, will be the primary dis-

air forces have been obliged to follow the lead of the most doctrin-ally up-to-date service, the US Army. Unlike the US Air Force, which has lately invested a great deal in its doctrinal renewal, the Canadian Air Force has still not put its doctrinal house in order.[7]

The primary US Air Force challenge to US Army doctrinal domi-nance in the late 1980s and early 1990s was an effects-based approach to operations based on John Warden's work, *The Air Campaign*.[8] One commentator described the US Air Force challenge this way: "The effects-based approach describes what effects are required to secure strategic objectives and then conduct military actions that would bring about the required effects. The US Air Force champions the effects-based approach and has developed it as a concept nested in a broader 'Rapid Decisive Operations' con-cept by Joint Forces Command."[9] An effects-based approach can be seen as an outcomes versus outputs approach to operations. For example, a recent MA thesis written at the US Army Command and General Staff College concluded that the US Army still uses an "objectives-based approach to operations" and recommends that it adopt an "effects-based approach to operations."[10] A detailed study of C[2] in the Gulf War found that senior commanders generally found it difficult during operations both to distinguish outputs from outcomes and to discover outcomes. In fact, the inability to discern what were perceived at the time to be outcomes (e.g., dam-age to specific enemy capabilities) was usually the reason senior commanders often focused on outputs (e.g., sortie rates), which did not necessarily have a direct bearing on the desired outcomes of the campaign.[11] Therefore, after the Gulf War, the US Air Force redou-bled its efforts to devise a truly effects-based approach to opera-tions. It should be noted here that unlike the Canadian Navy, with its unique networked-enabled capabilities, and the Army, with its distinctive approach to manoeuvre and operational art, the Cana-dian Air Force has not developed its own effects-based approach to operations, and it generally adheres to the US Air Force version of effects-based operations. The lack of originality in the Canadian Air Force's approach to effects-based operations is a result of the problems the Canadian Air Force has had in producing doctrine, as noted earlier, and in even properly documenting its operations.[12]

EFFECTS-BASED OPERATIONS The approach to operations champi-oned by the US Air Force, now formally known as Effects-Based

Operations (EBO), has become another buzzword in the current debate on how war and other operations should be conducted, and it is a term now used frequently in the joint arena.[13] A number of commentators have noted that EBO has its roots in ancient (Sun Tzu) and classical (Clausewitz) theories of wars.[14] However, the most recent branch on the EBO theory tree is one based on the writings of Warden and of Italian air power theorist Giulio Douhet. Douhet proposed solutions to the problems encountered by Western nations in the First World War, where stalemate at sea and on land caused widespread devastation and loss. He advocated a new style of warfare whereby aircraft would directly attack enemy vital centres, what might be called centres of gravity today, and bring future wars to a quick and decisive conclusion.[15] Ideas like these were modified or developed in parallel by airmen in the US and Britain to win or to maintain the "independence" of air forces from armies and navies from the 1920s through to the 1950s.[16] Therefore, Douhet's vision of EBO is the one most commonly used in air force circles; however, Ho notes that there is no authoritative definition of EBO, and he describes six different theoretical variants on the EBO theme.[17]

In general terms, EBO focuses on causal explanations to see whether actions that are planned or taken actually result in the desired effects. The key to achieving success with EBO is in predicting how physical actions can result in behavioural outcomes. In many ways EBO is a new way of describing an old concept because it has been at the heart of theories of air warfare since the earliest air power theorists, who were almost always concerned as much with the effects as with the means of applying air power. In fact, Douhet's theories were based on the notion of using the physical action of bombing to effect behavioural changes in the leadership of a nation. Critics of EBO have, therefore, used the failures of air power theorists in accurately predicting the outcomes (effects) of aerial bombardment to illustrate why true EBO may not be possible.[18] Some of these criticisms are based on the chaotic nature of warfare and the fact that chaos theory tells us that second- and third-order effects, especially those associated with human behaviour, cannot be predicted with the accuracy necessary to achieve the results EBO enthusiasts have claimed.[19]

While acknowledging non-combat aspects of EBO, some in the US Air Force still present it as largely a targeting exercise. For example, in an article purporting to represent the US Air Force approach to applying air power, Colonel Gary L. Crowder, the chief of strategy, concepts and doctrine of the US Air Force's Air Combat Command, focuses on the effects of new precision-guided munitions in executing EBO.[20] Those who favour this targeting approach to EBO have claimed that the initial "Shock and Awe" bombing campaign in Operation Iraqi Freedom (the US-led attack on Iraq starting in March 2003) was an example of Rapid Decisive Operations (RDO). The Shock and Awe concept comes from a 1996 paper, later published as a book, that was written by military strategists Harlan Ullman and James Wade titled "Shock and Awe: Achieving Rapid Dominance."[21] The theory appears to be very Douhetian in its concept of destroying the enemy will to resist by imposing "the nonnuclear equivalent of the impact of the atomic bombs dropped on" Japan, and very ambitious in its desire to "control the environment and to master all levels of an opponent's activities [so that] ... resistance would be seen as futile." To many this prescription seemed to fit the description of what was attempted by air forces in the early stages of Operation Iraqi Freedom. Ullman, however, stated that although the air campaign in Operation Iraqi Freedom "appears to come out of a book by strategic-air-power advocates, who have argued that you start at the center and work your way out to disrupt and destroy whatever," it was not what he envisaged as "shock and awe."[22] This example of different interpretations of the Shock and Awe concept demonstrates once again the problem with a number of current theories of war – they are, as noted earlier, still hazy, ill defined, and subject to different interpretations.

Critics of approaches to EBO that concentrate on targeting as a means of achieving outcomes caution that studying the theoretical foundations and historical examples of this type of EBO proves its futility as an approach to conducting operations. They note that strategic bombing theories like those of Douhet and Warden have underestimated the obstacles to achieving intended goals because attempts to destroy an enemy's will to resist by strategic bombing fail, unless much of his infrastructure is destroyed and his territory physically occupied, as was the case in the Second World War. As for the recent shock and awe variant of EBO theory, Kagan asserts that those who advocate this approach to warfare ignore the fact that the destruction of targets and resultant killing of civilians necessary to achieve the desired effect may undermine the political objectives of the campaign.[23] The challenge for champions of EBO will be to see whether modern theories, methods of analysis, and technology can make true EBO possible.[24]

A number of advocates of NCW have recently portrayed EBO as an adjunct to the theory of NCW; however, proponents of EBO would argue that EBO focuses on outcomes more than NCW, which focuses on inputs, i.e., the network. For proponents of EBO, networks are enablers for EBO and should not be seen as the primary consideration in devising new ways of war and other operations.

Whatever their differences, proponents of both EBO and NCW have focused on the technical rather than the human dimension of war. Many commentators have identified the need for more attention to be paid to the human dimension of EBO, but the complexity of this effort has been equated to "Ph.D. level warfare."[25] However, as with NCW, confusion over what EBO really means has led to a situation where "the concept is neither thoroughly nor evenly understood among military people," and as a result, "[o]nly now is EBO being tentatively and unevenly incorporated into service and joint doctrine."[26] Furthermore, not everyone has accepted EBO as a doctrinal concept. In September 2005, the US Marine Corps formally repudiated EBO as a philosophy of war largely because they object to EBO's focus on technology and its centralized command philosophy. The US Marine Corps argues that EBO is incompatible with its doctrine of mission command-driven, human-centric operations.[27] As this example shows, until a fully developed theory of EBO is validated and accepted, it will be an uncertain guide for transformation initiatives.

Once again, we see that a number of "theories" of war, such as EBO, RDO, and Shock and Awe, are evolving concepts that should be used carefully and subjected to more debate and research before they are accepted wholeheartedly as the foundation of any major changes to armed forces.

NETWORKS AND AIR FORCES While Western navies focused on connecting their ships more closely using electronic networks at the end of the twentieth century, air forces had already achieved this type of networking, especially in air defence operations. The earliest networked system, in a modern sense, was arguably the British air defences developed during the First World War to counter the attacks of first German Zeppelins and then fixed wing bombers.[28]

Brigadier General E.B. Ashmore commanded an integrated air defence system that comprised an observer corps, searchlights, anti-aircraft artillery, and fighters, all linked by a sophisticated communications network (by First World War standards) that permitted control centres to coordinate the activities of the system. By November 1918, this system involved 20,000 personnel, more than 500 guns, 600 searchlights, and 16 fighter squadrons.[29]

By the standards of the day, this was a highly innovative network. Early warning was improved by putting huge double discs and concrete mirrors into the channel cliffs to gather the sound from approaching bombers, and with practice operators could reportedly locate the bearing of approaching enemy aircraft up to twenty-four kilometres away. Emphasis was also placed on using directional sound devices for aiming searchlights and guns.[30] In less than two years the British had developed a system that was quite effective in providing a common operating picture, according to Ashmore's own account:

> I sat overlooking the map from a raised gallery. In effect, I could follow the course of all aircraft flying over the country as the counters crept across the map. The system worked very rapidly. From the time an observer at one of the stations in the country saw a machine over him, to the time when the counter representing it appeared on the map, was not, as a rule, more than half a minute. In front of me a row of switches enabled me to cut into the plotters' lines, and talk to any subordinate commanders at the sub-controls.[31]

Ashmore's system was the template for subsequent air defence organizations which were improved by a radar and air-to-ground radio. The British air defence system used in the Battle of Britain, the German air defence system deployed after 1940 to protect continental Europe, and the North American air defence system built during the Cold War were all highly networked systems along the Ashmore model.

However, from a theoretical perspective, air forces preferred the offence to the defence, and as we have seen, they have preferred to concentrate their theoretical attention not on defensive networked systems but on the outcomes that can be achieved by strategic attack. Although guilty in the past of focusing their force structure planning on technology, platforms, and inputs,[32] air forces now favour outcomes-based theories like EBO that fit better into the air force portion of joint operations. This new focus was articulated recently in an essay in the *RAF Air Power Review*, which stated that the "key to the synergy of the joint force" is EBO, and the mechanism to achieve that synergy is the Air Operations Centre.[33] For air forces, then, networks are a necessary enabler, but they are secondary to the main focus, which is what is achieved and not how it is achieved. For some, focusing on networks rather than effects is a step backward along a road of conceptual development based on EBO.

Another problem with the current theory of NCW from an air force perspective is its emphasis on self-synchronization and mission command or command-by-influence.[34] Synchronization as a concept of operations is emphasized more by land forces than air forces. In comparing US Air Force and US Army doctrine, it can be seen that the US Air Force focuses on the integration of air power across the entire joint theatre of operations, whereas the US Army tends to think geographically and emphasizes the synchronization of actions in time and space. It has been argued that the Army approach contrasts with the more holistic US Air Force perspective, which focuses on the effects that massing forces through integration can achieve.[35]

In an NCW context, Roddy notes that Cebrowski originally defined self-synchronization as "'the ability of a well informed force to organize and synchronize complex warfare activities from the bottom up.'" He also notes that more recently it has been suggested that "self-synchronization 'calls for lower-level decision makers to be guided only by their training, understanding of the commander's intent, and their awareness of the situation in relevant portions of the battlespace,'" and that "'[s]elf-synchronization emerges when units within a force use common information, the commander's intent, and a common rule set – or doctrine – to self-organize and accomplish the commander's objectives.'"[36]

At fairly low tactical levels, when close coordination among many air assets is not essential and threat levels are low, self-synchronization and command-by-influence can by employed by air forces; however, in other circumstances these processes can be problematic. For example, when decisions have enormous political consequences, such as the release of nuclear weapons or shooting down civilian aircraft, decision making will be retained at the highest levels, and one would be hard pressed to imagine a plausible scenario where these types of decisions would be susceptible to self-synchronization or command-by-influence processes.[37] A recent *Joint Force Quarterly* article put it this way: "Because of casual linkages among target sets and the danger of objective fratricide, effects based operations must be orchestrated by a centralized planning and execution authority that has situational understanding of every aspect of the diplomatic, informational, economic, and military campaign."[38]

In other circumstances, such as when large air forces need to conduct operations against an enemy with some credible air defence capability, neither self-synchronization nor command-by-influence are likely to be of much use except during short periods of time at the lowest tactical levels. For example, in Operation Allied Force, an air campaign against a very weak state but one with some air defence capability, complicated command and control arrangements were necessary to coordinate the activities of hundreds of air assets down to the minute (or less). The idea of allowing the vast number of air assets involved in such operations to self-synchronize or to use command-by-influence is difficult to imagine. One author notes that to achieve unity of effort, "the realities of modern joint air operations ... require centralized planning and direction" at "the highest levels."[39] Crowder tells us that a critical element in achieving unity of effort while executing EBO, from an air force perspective, is the Air Tasking Order, which provides "a common command and control architecture for all the air players that are involved."[40] The nature of complex air operations suggests that while there may be limited opportunities for self-synchronization and command-by-influence processes, for the foreseeable future air forces will rely on command-by-plan to execute their missions. There are, therefore, unique aspects to employing air power that make NCW's emphasis on synchronization and mission command inappropriate from an air force point of view.

If the principle of self-synchronization seems difficult to apply to air forces, dependent as they are on command-by-plan as represented by Air Tasking Orders and produced by discrete organizational structures as was done for Operation Iraqi Freedom, the idea of a self-organizing system, as proposed by FORCEnet 21,[41] seems almost beyond the realm of plausibility.

Therefore, air forces today and in the foreseeable future rely on command-by-plan and, in certain cases, such as when a command decision could have important political repercussions, even command-by-direction. While air force c^2 organizations and related joint organizations depend on networks to accomplish their tasks,[42] the network is not the focus; it merely enables the activity – EBO.

5

The Army Paradigm: Manoeuvre and Operational Art

After its traumatic experience in Vietnam, the US Army searched for a new focus for its approach to war. At the same time that the US Navy was developing networked operations and the US Air Force was debating the merits of EBO, the US Army modified the European concept of operational art in a "renaissance" of this theory of war. Commonly defined as "the use of military forces to achieve strategic goals through the design, organization, integration, and conduct of theater strategies, campaigns, major operations, and battles,"[1] most Western doctrine advocates the application of operational art to allow military professionals to orchestrate campaigns that link tactical actions with strategic objectives. Until Cebrowski introduced NCW as the prevailing concept in US military transformation at the beginning of the twenty-first century, operational art was in many ways the dominant paradigm in US military thought. The notion of operational art was the driving force behind the creation of the US worldwide regional command system – whose regional commands were originally referred to as "C-in-Cdoms"[2] because of their immense influence in their respective parts of the world – and remains the foundation for most Western joint doctrine. To counter the land-centric focus of the operational art concept, air forces and navies challenged many of the US Army interpretations of operational art and devised their own concepts to explain how they might practice it.

This chapter examines the development of operational art and associated ideas, like manoeuvre, in the US Army. It then explores navy and air force challenges to the US Army interpretation of the concept of operational art; some possible future developments of the concept; Canadian variations on the US Army concept; modern theories of manoeuvre; manoeuvre in a Canadian context; manoeuvre and the contemporary operational environment; manoeuvre, technology, and doctrine; and networks and the Canadian Army. The chapter concludes with an examination of the emerging Canadian paradigm of networked operations.

AMERICAN ARMY RENAISSANCE The US Army has had a disproportionate influence, until recently, on the development of joint doctrine, particularly at the operational level. A key person in the creation of modern US Army doctrine was General Donn Starry, who, according to his most avid admirers, was almost exclusively responsible for the transition of US military thought from the "technical shallowness of an incoherent tactical doctrine to an advanced operational consciousness." Furthermore, by institutionalizing scientific patterns of research, criticism, and constant change, he determined the dynamic nature of American military thought for the future.[3] To establish the context for the evolution of the operational level of war at the end of the twentieth century, a brief summary follows of the development of American military thought according to Starry, who as commandant of the US Army Training and Doctrine Command (1977–81) at the height of the renaissance in American military thought had a considerable impact on joint doctrine today.[4]

According to Starry, up to 1945 (with very few exceptions), the US military system was designed to overwhelm potential enemies by mass in a battle of military and national annihilation using the production techniques of the Industrial Revolution. This approach was shared by many of America's allies at the time. As Canadian historian Bill McAndrew put it, Allied commanders in the Second World War framed their campaigns on the attritional model of 1914–18.[5] They were inclined to use technical means to meet operational problems and usually attacked an opponent at his strongest point "after burying him with tons of high explosive."[6]

Immediately after the Second World War, according to Starry, the offence was portrayed as the dominant form of combat in American military thought, and the Jominian precept of massing at the decisive point was the preferred operational technique. With the advent of the Cold War, however, the US Army could no longer be assured of superiority of numbers against the Soviets in Europe, and various doctrines for the use of nuclear weapons (the so-called Pentomic Army, with nuclear warheads mated to almost every conceivable weapon; e.g., the Davy Crockett mortar) were put forward to redress the balance. Therefore, the operational level of war was neglected by both the Americans and the Soviets during the Cold War until the US became involved in a large hot war in Vietnam. Although the US Army believed that it had won the Vietnam war at the tactical level, this did not translate into strategic-level victory, and this experience was one of the major catalysts that precipitated the renaissance in American military thought of the 1970s and 1980s. The outcome of the Middle East War of 1973, with the importance of c^2, all-arms combat, the integration of technical advances (anti-armour missiles, precision-guided munitions, etc.) into war-fighting doctrine, and the fact the outcome of battle now rested on factors other than numbers, also helped to drive US Army doctrinal change.[7]

IMPACT OF THE AMERICAN RENAISSANCE The US Army school of thought has had a considerable impact on the practice of operational art today. Perhaps the most visible impact is its emphasis on manoeuvre warfare. The linking of manoeuvre on the battlefield with success at the operational level has sparked a lively debate among students of war.

Furthermore, by associating levels of war with geographical locations, like theatres of war, rather than with "categories of action," and by tying the concept of an "operational art" to one level the US Army imparted a distinctive, land-based flavour to these concepts. Despite the conceptual and semantic difficulties implicit in the US Army view, they have remained intact in US doctrine and have been incorporated in US joint doctrine since the original version of Joint Publication (JP) 3–0 in 1995.[8]

A further impact of the new US Army school of thought, which in fact constitutes the essence of the evolution of operational art in the US armed forces and the community of military theoreticians, is a shift from a paradigm of attrition by means of superior technology and tactics to one of advanced operational manoeuvre, according to Naveh. Even though Western armies have espoused some variant of manoeuvre doctrine since the end of the First World War,[9] he says that in the Gulf War the new operational art proved, for the first time in modern warfare, that the deterministic predisposition towards attrition, so common in Western military culture, had been replaced by a manoeuvre approach.[10]

Implicit in US Army operational-level doctrine is the belief that "Wars are won on the ground. Success or failure of the land battle typically equates to national success or failure. The culminating or decisive action of a war is most often conducted by land forces ... The application of military force on land is an action an adversary cannot ignore; it forces a decision."[11] This assertion that land forces are the pre-eminent weapon in the nation's arsenal relegates air and naval forces to a supporting role in the "decisive action of a war." This is still a contentious issue that will be examined in more detail shortly.

Perhaps the most enduring impact of the US Army's school of thought is its view that operational art "provides a framework to assist commanders in ordering their thoughts when designing campaigns and major operations. Without operational art, war would be a set of disconnected engagements with relative attrition the only measure of success."[12] This view has permeated virtually all joint doctrine, and the operational-level headquarters has become, for many, an indispensable adjunct to military operations. However, this view has been challenged recently as air forces and navies have articulated their doctrines more clearly.

NAVAL AND AIR FORCE CHALLENGES In response to the US Army's doctrinal dominance at the operational level, there has been the rise of what might be called heretical[13] challenges to joint doctrine founded on land-centric concepts. Grant puts it this way:

> Joint doctrine today carries forward a land-centric focus because it is still largely based on dominant surface maneuver. Key air concepts – and some naval concepts – receive short shrift. Differences between land and air components generally are resolved in favor of the land commander. Most of all, it is striking how closely joint doctrine runs parallel to the Army doctrine of maneuver, fires, and force protection. As a result, major conflicts in the joint-doctrine process most often erupt over differences between air and ground views of operational strategy, command, and organization.[14]

The air force challenge to the US Army's doctrinal dominance at the operational level has been discussed in the previous chapter on EBO.

Wylie presents the naval challenge to the concept of operational art with this argument: "The operational art is an artifice appropriate to ground force doctrine but the navy (and the air force) have no need for such a concept." In fact, navies have generally avoided the term "operational," instead preferring "doctrine" to indicate what lies between maritime strategy and tactics.[15] Hughes puts it somewhat differently, arguing that the "three prongs of the naval trident have long been called strategy, logistics and tactics."[16] Specific definitions aside, navies have traditionally seen doctrine in a different light than armies. Grant notes that for 200 years the US Navy has kept doctrine at arm's length for fear that a binding set of principles might restrict the initiative and independence of the captain at sea – the very foundation of naval combat. Therefore, strategy and tactics were the domains where naval officers concentrated their attention, and until recently, the bulk of US Navy doctrine was "found in the unwritten shared experiences of its officers."[17] But Desert Storm's joint-force air attack procedures jolted the US Navy out of its complacency, and it established a Naval Doctrine Command in 1993 in part to provide the doctrinal basis for its evolving statement of maritime strategy.[18] According to Tritten, "With the formation of the Naval Doctrine Command, the Navy now has its first centralized command responsible for the publication of doctrine for the fleet." But even with the formation of Naval Doctrine Command, basic US Navy doctrine is dated when compared with US Army and Air Force doctrine; the most recent version of Naval Doctrine Publication (NDP) 1 *Naval Warfare* was originally published in March 1994.[19] The naval approach to doctrine reflects its view of warfare at sea. Navies produce much less written doctrine than armies because of their view of doctrine as "a common cultural perspective of how the naval Services think about war ... and how they will act during time of war ... [therefore] Navy doctrine is the art of the admiral."[20] In theory, US Marine Corps doctrine is congruent with US Navy doctrine, but the Marines have generated their own interpretations of the US Navy's Operational Maneuver from the Sea and ship-to-objective manoeuvre (STOM) because some in the US Marine Corps believe that the US Navy has concentrated too much on maritime doctrine and neglected aspects of the land battle.[21]

Many of these concepts are not really applicable to medium powers such as Canada that do not possess forces capable of large-scale power projection. In a more nuanced fashion, the Canadian Navy has identified being a "joint enabler" as one of the basic principles of its naval strategy as laid down in *Leadmark: The Navy's Strategy for 2020.* In noting that "naval platforms and their crews are designed and trained for war-fighting at sea and in the littorals," it goes on to describe the synergistic effect of joint operations: "Navies cannot hold ground to the extent that an army can. Nor can they reach as swiftly to the far corners of the globe as an air force. But the ability of a navy to stand off a foreign shore for an indefinite period with substantial combat capability cannot be matched. Any joint ... concept of operations developed for the Canadian Forces must be undertaken in recognition of the unique attributes offered by each of the services."[22]

The heretical challenges of navies and air forces to the US Army's version of operational art are significant, but others have also questioned the current US army orthodoxy. The two brief critiques that follow are representative of other critiques.

THE END OF OPERATIONAL ART? Robert Leonhard has recently argued that the Jominian paradigm upon which current concepts of operational art are based is no longer valid. He contends that America's adversaries will not fight "campaigns of predictable and relatively short duration" based on geographical areas but will "prosecute unconventional campaigns that unfold over long periods of time" in disparate areas of the globe. In these circumstances, operational art as it is practised today will "wither away." As political, economic, cultural, and other factors "exceed the grasp and authority of regional combatant commanders and their staffs," campaign planning, "once easily confined to military operations in a given theatre," will become almost synonymous with strategy.[23] If Leonhard is right, then the expression "operational art" may be consigned to history's dustbin, and the term "operation" will revert to the meaning it had in the seventeenth and eighteenth centuries, when operations were an integral part of strategy.

FUTURE OPTIONS Colonel James K. Greer, the former director of the US Army's School of Advanced Military Studies (SAMS), argues that a new "operational-design construct" is required to address some of the challenges articulated above and to permit "the effective planning and execution of future campaigns and major operations." He describes five alternative avenues for redesigning US approaches to operational art. His second approach is based on the systems approach, which views all military organizations as complex systems, and it "would apply emerging systems and the science of chaos and the theory of complexity to developing an operational-design construct with which to execute the military equivalent of forcing opposing systems into either chaos or equilibrium." Of the five approaches, this is the most compatible with NCW, but in the operational art construct NCW would support the practice of operational art, not act as the organizing principle.[24]

THE CANADIAN SITUATION Despite the recent CF predilection to think of just about any military activity as having operational dimensions, Canadians have had very little experience in operational-level leadership roles. Arguably, the only Canadian to have held operational command, in the sense understood by American doctrine, is Admiral L.W. Murray in the northwest Atlantic area in the Second World War.[25] While some progress has been made since the mid-1990s, particularly in the Canadian Army, towards examining operational concepts from a theoretical and doctrinal point of view, there is no "sound intellectual base" in this country on which to base operational art. Rather, "a bureaucracy arbitrarily directed that operational art was to be adopted" from largely American sources.[26]

Coombs claims that this situation has caused a "fragmented" approach to operational thought in Canada, which explains why it does not always follow the tenets of prevailing Western doctrine. He also notes that Canadian use of operational art in peace support operations within the context of an alliance or coalition has strongly influenced its use by Canadians. Therefore, unlike many other militaries, the CF perception of the operational level of war is not focused on operational manoeuvre or operational logistics, nor is it tied to a theatre of war. Rather, Canadian commanders seek to coordinate operational-level systems appropriate to a multi-agency environment and the force structures under their command to achieve operational-level objectives. These ideas are discussed in greater detail later in this volume.[27]

Within that context, Gimblett has suggested that the Commodore Roger Girouard's command of task force 151 in Operation Apollo/ Enduring Freedom was a rare example of a Canadian exercising operational command in a coalition. As the commander of a task force of up to twenty ships from a dozen coalition nations operating in the Arabian Sea theatre, these naval officers could be considered operational commanders because they were assigned a clear geographical area of responsibility, commanded a relatively large force, and coordinated tactical actions that had strategic implications.[28]

Based on preliminary research into recent CF operations, it has been suggested that operational art may be practised by Canadian commanders if they exercise command in an area with clearly defined geographical boundaries; have the authority to employ forces within this area (this implies normally exercising operational command or operational control of the forces in the area); and undertake objectives directly linked to strategic aims. According to these criteria, the DCDS Group, despite claims to the contrary by some, does not exercise operational command in overseas taskings. Rather, for these taskings the DCDS is a "force deployer." In domestic operations the DCDS might be an operational commander, but it has been suggested that as an NDHQ organization, the DCDS Group really functions at the strategic level.[29] To address these issues, a major reform of the command and control of the Canadian Forces so that they more closely resemble the unified commands of the US armed forces was one of the objectives of the Defence Policy Statement unveiled in April 2005.[30]

MODERN THEORIES OF MANOEUVRE: THE OODA LOOP A key concept embedded in US Army operational art theory is that of manoeuvre. The fundamental assumption underlying operational plans is that they will be "manoeuvrist." Current concepts of manoeuvre embraced by the US Army and other services have been strongly influenced by a model developed by US Air Force Lieutenant-Colonel John Boyd. He retired in 1985 and died in 1997, had little in the way of command or combat experience, and based his model on observations of fighter pilots in training and in the Korean War. His OODA (observation-orientation-decision-action, or Observe, Orient, Decide, Act) loop model was designed to enable US forces to fight "smarter," employing mission-type orders (*auftragstaktik*) to effect a sort of "military judo" on the enemy by creating friction and exploiting enemy mistakes.[31]

Boyd's model, which has been referred to as the "Boyd theory," is not a novel concept, but according to some it is a synthesis of much of what has been written by past theorists of war. Others characterize it as a profound new theory of warfare. Boyd's model is simple but elegant. In it, every decision occurs in time-competitive OODA cycles. This process implies that military decision-makers need a psychological and temporal orientation instead of the usual physical and spatial orientation. There is a need for mental agility and creativity, comfort with ambiguity, and the confidence to allow subordinates to use their initiative. Boyd's model portrays the most important manoeuvres as taking place inside the enemy's mental processes (the enemy's OODA loop); therefore, the most important manoeuvre space is in the fourth dimension of time.[32]

Boyd's model calls for commanders and their staffs to constantly revise their mental models to stay inside an opponent's OODA loop. This process also has the effect of creating a mind-set more predisposed to fighting the enemy than fighting according to a pre-set plan, as is common with plan-based methods currently in use. Boyd's model is therefore congruent with pattern recognition theories of decision making, such as those of Gary Klein, that advocate naturalistic or intuitive decision making in time-sensitive situations.[33]

The OODA loop concept has been used by advocates of NCW to argue for increased speed in decision-making cycles or "speed of command."[34] However, as many critics have noted, faster decisions do not necessarily mean better decisions. As one critic put it, "The 'speed of command' characteristic of the NCW environment could lead to some undesirable effects. 'We may find ourselves acting so rapidly within our enemy's decision loop that we largely are prompting and responding to our own signals ... like Pavlov's dog ringing his own bell and wondering why he's salivating so much.'"[35]

Boyd never attempted to publish his ideas, but William S. Lind codified them in his *Maneuver Warfare Handbook*, which was specifically tailored for the US Marine Corps. Lind posited that future ground combat would be dominated by those who could decentralize actions and accept confusion and disorder while avoiding all patterns and formulas of predictive behaviour.[36] Proponents, like Lind, of manoeuvre based on the Boyd theory advocate a more dynamic approach to strategy and operational thinking than is currently found in some US military circles. Critics of the current system point out that the OODA model contrasts with the inherently analytical nature of US Army planning and decision making, which neglects the role of synthesis as an enabler of intuition in the Boyd theory. This has caused rifts in the US Army, where some advocate radical and bold culture shifts to allow for true mission command, while others suggest that the present model of centralized planning and decentralized execution is sufficient to meet future needs. Polk argues that true manoeuvre warfare, as described by the Boyd model, cannot be practised by the US Army because the use of initiative and the toleration of mistakes are antithetical to US Army culture today. This may not bode well for the future because in a culture where conformity is rewarded more than initiative, those who rise to command are being selected on criteria that will not allow them to be proficient practitioners of operational art as Boyd envisioned it.[37]

This situation is exacerbated, according to Polk, because one of Boyd's most important insights, his emphasis on the importance of time, has been lost in a doctrinal "dumbing down process." Too often, Polk claims, the OODA loop process is portrayed as one of making decisions more quickly than the enemy. But "out-OODAing" an enemy is more a process of achieving temporal effects than just being faster (or slower) than an enemy in decision making. Fadok argues that Boyd's approach is predominantly Clausewitzian because manoeuvring inside the enemy's mental processes as depicted by the OODA loop is a more philosophical, abstract, and nonlinear approach than the approach advocated by Warden. In other words, Boyd's theory is about "err-power" – how to make the enemy lose versus how to win ourselves.[38]

Most theorists of manoeuvre agree on the *ends* of manoeuvre warfare: to defeat the enemy quickly, decisively, and with minimum loss. The *means* of achieving these ends, however, are varied and depend largely on which war-fighting community the "manoeuvrist" comes from. Theorists tend to focus on the means they know best, and true joint manoeuvre theory is handicapped by the largely single-service approach taken by the US services. Furthermore, support and logistics, for some the heart of operational art, are often overlooked in manoeuvre theory.

Few would argue that manoeuvre is not a necessary part of operational art, but a number of commentators remind us that manoeuvre today has been suggested as a solution for problems that are beyond its capacity to solve. First, it is often portrayed as a solution for the perceived predisposition for casualty aversion in the West, when in fact manoeuvre warfare between roughly equal opponents (e.g., the last "Hundred Days" battles of the First World War and the Eastern Front in the Second World War) has resulted in very high casualty rates indeed.[39] Second, it has become a mantra for some that automatically excludes other possibilities for fighting, like defensive attrition, which are then not fully explored when operational plans are devised. In today's climate of doctrinal flux, perhaps it is best to keep an open mind. Remember that the word "manoeuvre" conjures up many possibilities in different war-fighting communities, but that a great deal more study is required before all of its possibilities are clearly understood.

MANOEUVRE IN A CANADIAN CONTEXT The legacy of the Canadian Army at the dawn of the twenty-first century is comparable to that of many Western nations. It was a small, professional colonial army at the end of the nineteenth century, which provided the core of national mobilization in the First and Second World Wars. It participated in the Korean War, and since the 1950s it has been fully engaged with peace support operations.[40] The Army weathered the lean inter-war years of the 1920s and 1930s, as well as expanding and contracting to fulfil national requirements during the Cold War and post-Cold War eras. The doctrinal paradigm of Canada's land forces over the last hundred years has migrated from attrition to manoeuvre to reflect that history, and the Canadian Army has become the leading proponent of manoeuvre in the CF. Using Lind's precepts, it has articulated its vision of manoeuvre warfare for land operations in its current keystone doctrine manual of April 1998, *Canada's Army: We Stand on Guard for Thee*. For example, it provides a key tenet of manoeuvre warfare, comprehension of the higher commander's intent, as a principal component of command:[41]

> The principle of subsidiarity is to be applied. Subordinate commanders are to be given, to the greatest extent possible, the responsibility, information, and resources to act as the tactical situation demands, without further reference to higher authority. In effect subordinates are empowered to perform and respond to situations as their commander would have, had their commanders been there in person. To realize this command philosophy, leaders must know their subordinates intimately and

trust them implicitly; subordinates in turn, must not only be skilled in the military art, but fully aware of their responsibilities to their commander and committed to fulfilling them.[42]

This philosophy is predicated on a particular state of mind or manner of thinking rather than on techniques and procedures.[43] It is an attitude that strives to defeat an adversary by destroying his source of moral, cybernetic, or physical power.[44] The objective, or end state, of the manoeuvrist approach is to negate the enemy's ability to conduct warfare as a cohesive force.[45] In this vision, manoeuvre warfare should focus on enemy vulnerabilities, not ground; avoid enemy strength and attack his weaknesses; concentrate on the main effort; and be agile. To achieve these ends the manoeuvrist commander should support manoeuvre with fire, exploit tactical opportunities, act boldly and decisively, avoid set rules and patterns, use mission-type orders, and command from the front.[46] These ideas were further affirmed in *Conduct of Land Operations – Operational Level Doctrine for the Canadian Army*, issued in July 1998.[47]

The Canadian Army's concept of manoeuvre emphasizes that the defeat of the enemy can best be achieved by "bringing about the systematic destruction of the enemy's ability to react to changing situations, destruction of his combat cohesion and, most important, destruction of his will to fight." Nevertheless, Canadian Army doctrine recognizes that "attrition may not only be unavoidable, it may be desirable," depending upon "the commander's intent for battle." The use of operational art by land forces in Canada is founded on the command philosophy of what they call "trust leadership." Using this philosophy, commanders at all levels are expected "to issue mission orders along with their intent and then allow their subordinates to get on with their tasks." However, it is recognized that this philosophy may be difficult to achieve in practice, "since it is inherent to the nature of the military to over-control its subordinates, and with modern information and communication facilities, it is becoming increasingly easy to do so." Canadian Army doctrine cautions us not to confuse the concept of manoeuvre warfare with manoeuvre. While manoeuvre is defined as "the *employment of forces through movement in combination with speed, firepower, or fire potential, to achieve a position of advantage in respect to the enemy in order to achieve the mission*" [emphasis in original], manoeuvre warfare is described as "a mind set." Canadian Army doctrine goes on to say that "There are no checklists or tactical manuals that offer a prescribed formula on how to employ manoeuvre warfare. Leaders at all levels must first understand what is required to accomplish a superior's mission and then do their utmost to work within the parameters set out for that mission." It concludes by describing manoeuvre warfare as "an attitude of mind; commanders think and react faster than their foes in order to mass friendly strengths against enemy weaknesses to attack his vulnerabilities be they moral or physical."[48]

Proponents of manoeuvre believe the physical conduct of military operations to be circuitous. They visualize the efforts of military forces as being directed towards the creation, exploitation, and enhancement of misdirection rather than force-on-force confrontation. Manoeuvrist military activities will disorient, disrupt, and strain enemy systems to the breaking point by attacking indirectly their centres of gravity.[49] In essence, the tempo of operations must be such that the enemy is forced to conform to our plans, to the point where he can no longer react in a coherent manner. Moreover, the concept of directive control is utilized to provide a philosophy of command.[50]

Directive control, sometimes known as directive command, uses higher commanders' intent, mission analysis, and designation of a main effort to promote rapid manoeuvre in the physical and conceptual sense.[51] Effective implementation of this method of command is contingent upon the decentralization of authority. In conjunction with awareness of the larger purpose of tasks,[52] this permits subordinates to implement operations that employ rapid tempo and synergistic effects to achieve decisive results.[53] The differences between command in attrition and command in manoeuvre warfare are simplistically depicted in Figure 5.1.[54]

MANOEUVRE AND THE CONTEMPORARY OPERATIONAL ENVIRONMENT Manoeuvre doctrine enables a small professional army to conduct operations across the spectrum of conflict, from peace support and related activities to kinetic operations.[55] It supplies commanders with the flexibility to use military power in a manner designed to effectively utilize limited resources in most types of situations.[56] Nonetheless, it has been recognized in recent years that to be successful, practitioners of manoeuvre philosophy had to place much greater emphasis on the creation of shared awareness to

Attrition warfare	Manoeuvre warfare
Physical	Psychological
Positional	Fluid
Centralized Authority	Decentralized Authority

Figure 5.1 Parameters of command based on concepts in Simpkin, *Race to the Swift*, 206.

reduce the ever present ambiguity and corresponding friction of current operational environments.

This recognition has come from the experiences of modern peace operations. These military activities involve comprehensive campaigns that simultaneously address diplomatic, informational, military, and economic aspects of the environment. At the same time the setting is asymmetric and usually non-permissive, with innumerable state and non-state actors.[57] Land operations are complex and continuous, regularly involving physical and psychological isolation and more often than not, until too late, unseen lethality.[58] Adversaries are often non-state and motivated by issues other than that of policy. They attack in unpredictable ways, using the strengths of an opponent as a weakness to gain a temporary advantage that can be exploited. Conflict is not confined to discernible regions, and all aspects of the spectrum of conflict are involved in a specific area, with distinctions between combatants and non-combatants disappearing. Effectiveness in joint, multinational, and multi-agency operations is the key determinant of success.[59] Furthermore, because of the increasing role of technology, the intellectual dimension of warfare has increased. It is no longer viewed as a clash of wills between two opposing commanders but as a contest of thinking, interconnected adversaries, each trying to triumph over the other.[60] For the Canadian Army the military aspects of such warfare are a prescription for manoeuvre philosophy and require mature, experienced leaders, in addition to cohesive units that are capable of independent operations.[61] The summary of conclusions from the CF *Debrief the Leaders Project (Officers)* reinforces these ideas and indicates that officer professional development will need to emphasize critical thinking, strategic conceptualization, and effective decision making, as well as the ability to understand and work in diverse cultures:

- While the prime function of the CF remains the application of military force in support of government policy, the use of force will be in discrete amounts fully integrated with, and usually subordinated to political, diplomatic and economic measures.
- The need for global security will continue to place a great premium on leadership in the future, but new competencies are needed to supplement traditional leadership competencies as defined by another era of war and fighting.
- Strategic and operational knowledge and skill sets must be created, over and above the excellent tactical training that historically has characterized the CF.[62]

MANOEUVRE, TECHNOLOGY, AND DOCTRINE To function in modern operational settings, the Canadian Army must change from a hierarchical, centralized organization to a vertically and horizontally networked information-age structure. This idea was presaged by the 1994 White Paper, which outlined the necessity to augment command and control structures and decrease the number and size of headquarters to streamline communications within the CF.[63] At that time it was stressed that technological enablers were required to provide troops with "the means to carry out their missions."[64] In 1996 the commander of Land Force Command further elaborated these requirements in the Land Force Information System Project, which was designed to increase the capacity of command, control, communications, and intelligence at the operational and tactical levels: "Success in future operations demands the ability to execute one's own decision cycle within that of any opposition."[65] In this sense technology would be used to actualize a key principle of manoeuvre warfare doctrine.

It is significant that such ideas concerning the role of technology in land operations were articulated in Canadian Army doctrine in a form that one can recognize today as a rudimentary description of NCW predating the work of Cebrowski and Garstka. The Army doctrine manual *Conduct of Land Operations* emphasized that developments in technology, particularly advances in the ability to capture, process, and disseminate information, coupled with the increased precision and lethality of weapons systems, had created a need to ensure that military forces were dispersed and interconnected. Ideas of shared awareness and decision making at the lowest possible levels were also expressed in this document:

Technological advances in all areas of science continue to change the face of conflict. The accuracy, lethality and range of modern weapon systems have forced commanders to disperse their formations, and therefore decentralize decision-making and execution. Also, improved surveillance and target acquisition from space and aerial platforms has decreased the freedom to manoeuvre. The speed of decision-making, the synchronization and concentration of force have increased in importance and all depend on accurate information. Meanwhile, technologies such as digitization have increased the ability to share information so that friendly and enemy forces will be more dependent on the electromagnetic spectrum to obtain and transmit information. As a result, our own vulnerabilities must be protected and the enemy's exploited.[66]

These thoughts concerning the collection and dissemination of information and intelligence products had been previously articulated in a rudimentary fashion in *Canada's Army* and were further developed in the 1999 doctrine manual *Land Force Information Operations*. The latter refined concepts initially contained in the *Conduct of Land Operations* concerning the role of information technology in the establishment of connectivity between a myriad of dispersed friendly entities. These relationships would create information superiority within the linked group to permit coordinated and decentralized operations against potentially larger threat forces. The term "network" was used in the context of the hardware required to enable the battlespace visualization necessary for victory in such an environment.[67]

Concurrently, other organizations besides the Directorate of Army Doctrine (DAD) within the then nascent Land Force Doctrine and Training System were examining ideas concerning the potential impact of technology on various aspects of the Army.[68] The Directorate of Land Strategic Concepts studied ideas of networks from a human and technical perspective in March 1999. This introspection was in aid of the development of future Army capabilities, not NCW per se. However, it was acknowledged that research was needed into the potential impact of battlefield digitization.[69] Later that year the Directorate of Land Strategic Concepts published *The Future Security Environment,* which cursorily mentioned digitization initiatives in allied armies but did not address concepts of NCW.[70] However, by

1999 the directorate was experimenting with the information technology architecture that would be necessary to support Intelligence, Surveillance, Target Acquisition, and Reconnaissance (ISTAR) capabilities. This information technology architecture mirrored some of the ideas discussed by Cebrowski and Garstka, predominantly the notions of interconnected sensors providing information that would be processed or "fused" and that would send the resultant product to "shooters," who would engage the opposing forces. This study also called for a change in doctrine and organizations to make best use of the new technology. Conspicuously, the human dimension of these networks was also emphasized: "The aim of this experiment was to identify more than technologies; it dealt with the whole of the human-machine system."[71]

These themes were echoed in early articles discussing network-centric capabilities within the CF, notably "The Canadian Forces and the Revolution in Military Affairs," by Vice-Admiral Gary Garnett, and "Command in a Network-Centric War," by Colonel Pierre Forgues. These commentaries discussed the network-centric environment in which it was believed that war would be fought in the twenty-first century. They argued that although technological enablers would not change the nature of command, they would greatly enhance the speed of decisions and subsequent action, as well as produce greater time for the formulation of appropriate actions and reactions. The ability to share intent and concepts throughout the network simultaneously would, it was thought, greatly enhance the commander's ability to empower his subordinates to make accurate and timely decisions. This ability to self-synchronize would then permit commanders to establish and maintain unity of effort with less friction than in previous epochs.[72]

The directorate's monograph *Command, Sense, Act, Shield and Sustain* appeared in 2001, shortly after these commentaries. This work used "digitization," a term first used by the US Army, to combine the operational framework of Command, Sense, Act, Shield, and Sustain into an integrated, complex system.[73] First, Command was the locus of all functions and integrated them towards specific goals. Second, Sense united sensor and sensor analysis capabilities into a single concept to eliminate "stovepiping" of information and enable knowledge networks. Third, Act united manoeuvre, firepower, and offensive information operations to create the desired effect at the tactical, operational, or strategic levels and was relevant to all types of operations. Fourth, Shield encapsulated defence on the moral and physical planes. Last, Sustain supported the moral and physical planes.[74]

These five land force functions had been introduced in 1999, as the then eleven combat functions were considered to be tactically oriented and did not address the moral domain of conflict. In order to develop a paradigm based on manoeuvre doctrine, it became necessary to define functions that would not only incorporate the moral and the physical planes of conflict but also be integrated across the levels of war. It was also proposed that information was the medium in which these functions would operate and would be a common factor to all. Technology would permit changes that would have far-reaching impacts on the organization and employment of the Army of the future; however, it would not supplant humans or the nature of command. Digitization would permit the Army to become an efficient practitioner of manoeuvre warfare by increasing the rapidity with which information was sensed and processed, thus enabling decisions to be swiftly made and enacted across the levels of war.[75]

Ideas of "network centricity" were expressed by the Directorate of Land Strategic Concepts in 2001 as a result of future force wargaming experiments that incorporated the new operational systems. It was noted in the experimentation report that there had been questions involving the integration of network-centric concepts and command, but it was believed that in the final analysis the war gaming demonstrated that network centricity would allow for both mission and directive command.[76] These relationships between humans, doctrine, and technology were reaffirmed in succeeding experimentation by other organizations. For example, the results of the Land Force Command and Control Information System (LFC²IS) evaluation in 2002 and of collaborative planning experimentation in 2004 supported these ideas.[77] In 2004, these premises coalesced in *The Force Employment Concept for the Army*, which linked manoeuvre warfare and network-enabled operations with future forms of EBO.[78]

NETWORKS AND THE CANADIAN ARMY Concepts concerning the role of technology in facilitating manoeuvre warfare have been expressed in various fashions by the Canadian Army since the 1990s and indicate a desire to understand the ramifications of the information age. Terms such as digitization, NCW, network-centric operations, Network-Enabled Capability, and NEOps have all been used to define this amorphous relationship between war and emerging technology as well as implications for doctrine.[79]

It is evident in this discourse that the role of technology, which is prominent, has been to actualize the relevant manoeuvre doctrine. For Canada's Army, ideas of NCW or NEOps constitute neither a fundamental change in the manner in which wars are fought nor an RMA.[80] McAndrew supports this contention in a study of the tension between technology and the human factor since the First World War, in which he demonstrates that the introduction of new technology has not normally led to a complete change in the nature of war, but rather to an accommodation between the newly adopted equipment and its military user. He puts forward the idea that even though the information age has led to new challenges, the Canadian Army, as in previous historical periods, must ensure this accommodation between technology and the soldier.[81]

Questions concerning the existence of an RMA and the necessity of establishing an interface between technology and people notwithstanding, the primacy of doctrine in any efforts by the Canadian Army to adopt and implement NEOps cannot be understated. McAndrew indicates the importance of doctrine to the Canadian Army through an examination of recent historical experience:

> While Canadian commanders lacked the need, means, and inclination to think or function at the operational level in the Second World War, they certainly felt the effects of others' use and misuse of its concept. This is particularly evident at the junction at the top with strategy, from which Canadians were excluded, and at the bottom with tactics, with which they were very much concerned. The point between the operational and tactical levels would seem to be a vital factor in military effectiveness. Mere tactics themselves may not win wars, but the purest operational conception will remain barren if the tactical means to implement it are deficient. Several factors link the tactical and operational levels, for example, troop quality, organizations and technology. The most vital ingredient, however would seem to be doctrine; not doctrine as dogma, but simply the shared premises, assumptions, and procedures that allowed soldiers, units and formations to function as a coherent whole.[82]

In an attempt to define these connections between warfare, technology, and doctrine, a number of theorists have promulgated concepts that seem to be applicable within the context of the Canadian Army, ideas of manoeuvre warfare, and the human dimension of NEOps. Martin van Creveld describes command as a search for certainty, in which a commander predicates all decisions on having information regarding all factors pertaining to a particular dilemma before making a choice.[83] For manoeuvrists this creates tension between the seemingly exclusive imperatives of doctrine and command. Van Creveld describes this conflict simplistically: "The history of command can thus be understood in terms of a race between the demand for information and the ability of command systems to meet it."[84]

Van Creveld hypothesizes that the greatest danger of increased information technology within information-driven command systems is that of distinguishing the relevant from the masses of the information available.[85] The danger of this is borne out by a report from recent American tactical operations in Iraq:

> Making every soldier a sensor briefs well but when it comes to execution there are many hurdles to overcome before we get there ...
>
> There is no staff at the battalion or brigade level capable of managing the information load we would generate at the 75% solution. Can't say how many patrols out of the total make it to brigades and can't imagine how someone could accurately generate that statistic, but would offer that not all data is important and tying up bandwidth burying the analysts under an avalanche of raw information is counter-productive at best. No staffs are manned to collect and process everything that comes in from the field right now, and analyzing what does come in effectively is close to impossible. That said, it is important to do the best one can to sort the wheat from the chaff. Platoon leaders and company commanders do the best processing and analysis with their own heads. Periodic assessments on their part submitted higher provide a much clearer picture than 60 patrol debriefs a week. Reports help with pattern analysis at higher HQ [headquarters], but most of what comes out of that analysis we already know.

... Human connectivity is every bit as important as digital connectivity.[86]

Van Creveld reinforces the importance of people by focusing on the human aspects of command as paramount in overcoming the inherent friction of war and winning the conflict of opposing wills. With this in mind, one must view command using manoeuvre doctrine in NEOps as an exercise in the art of leadership. Richard Simpkin articulates this idea simply, as a "supple chain": "a chain of trust and mutual respect running unbroken between theatre or army commander and tank or section commanders."[87] By using technology to assist with articulating the commander's intent, conducting mission analysis, and designating and using a main effort, the decentralization of command will occur. In this setting, subordinate commanders will make appropriate decisions and take action to achieve positive results without specific orders.[88] Van Creveld suggests that organizations can design their command structures to operate in the environment of chaos, or the "province of uncertainty," where less information can actually increase their likelihood of success. He outlines five principles for organizing command systems to achieve success, all of which are reflected in various current works concerning the Canadian Army: first, decision-making authority should be as low in the chain of command as possible to promote freedom of action; second, organizations should make decentralization of decision making possible by structuring units, at the lowest level, to be capable of self-sufficiency in operations; third, reporting and information systems need to work reciprocally throughout the organization; fourth, headquarters must not only rely on units to send information but maintain an active search capability outside headquarters to supplement this information; and last, formal and informal networks of communications must be maintained.[89]

These theoretical ideas are borne out by practice as doctrine, future studies, and experimentation support these contentions and others. It is obvious that the human factor is paramount, and the networks established by the Canadian Army seem to reflect this truth in that they are hybrids of both technology and people. In this situation command is predicated on communication, dissemination of intent, creation of shared awareness, and decentralized decision making.

This vision of war is not the result of an information age RMA or a particular arrangement of information architecture but the legacy of the historical past. The Canadian Army professes to be a doctrine-based organization that uses technology to increase its capability to practice manoeuvre warfare. Attempts to grapple with NCW-type concepts have been noticeable since the early 1990s, before the publication of Cebrowski and Garstka's work, and they have developed in accordance with many influences, in particular the need to use emerging technology in a way that will accommodate the user and actualize manoeuvre doctrine to permit a small professional army to operate in the contemporary environment of conflict. These themes will be developed further shortly.

THE EMERGING CANADIAN PARADIGM

While the idea of networked forces has been present in CF debates about change for some time, only in the past year or so has NEOps emerged as an overarching concept to drive CF transformation. For example, as recently as November 2003, Department of National Defence (DND) transformation was described in terms of three separate initiatives based on transforming the Navy, the Army, and the Air Force separately, as these excerpts from the National Defence Strategic Capability Plan indicated:

> Defence [DND] envisions the continued transformation of our Maritime Forces through the operational development and preservation of a globally deployable and network-centric Naval Task Group capable of sustained effort in combat and stabilization operations ... Defence envisions the Land Forces transforming from an army doctrinally based on manoeuvre supported by mass and area suppression firepower to precision effects based firepower supported by manoeuvre ... Defence envisions transforming the Air Force into a networked force capable of seamless operations with joint and/or combined forces at home or abroad.[90]

There is no mention in this document of NEOps; therefore, the Canadian concept of NEOps that is emerging from the experience of both the Navy and the Army is still in its early stages of development.

CURRENT CANADIAN VIEWS OF NEOPS Although NEOps has not yet been formally adopted as a concept in support of transformation, it has received significant attention among members of the Canadian defence community.[91] A good summary of current Canadian views on this evolving concept is contained in a draft report titled "Network Enabled Operations: A Canadian Perspective." Some of the key points of this report are summarized here. The report notes that NEOps is described in official Canadian strategic documents as "'a concept that has the potential to generate increased combat power by networking sensors, decision makers and combatants to achieve shared battlespace awareness, increased speed of command, higher operational tempo, greater lethality, increased survivability, and greater adaptability through rapid feedback loops.'"[92] The report also notes that while accepting the basic tenets of networked operations as described in the American literature, the Canadian approach to NEOps changes the focus from technology as the basis for networked operations to the human elements of these types of processes. This is congruent with the idea commonly shared by many Canadian commentators on NEOps that "whatever definition of NEOps Canada adopts, it should be consistent with Canadian culture and ethos" and that "Canada should be careful not to simply adopt a conceptual structure from the US that may not be consistent with the nuances and priorities" of Canadian military and other government agencies. This implies "that the concept of NEOps requires more than simply overlaying a networking capability" onto an existing organizational or command and control structure. A number of Canadian commentators on NEOps believe that "adopting NEOps in Canada would require a core paradigmatic change in the military as an organization, and the reworking of its relationships with all other members of the network." Perhaps most importantly, from a Canadian point of view, using NEOps in "the JIMP [Joint, Interagency, Multinational, and Public] context requires the recognition that networks extend beyond the mere use of information technology and into the realm of social networks." Therefore, introducing NEOps will require transformational activities to effect the necessary paradigm shifts.[93]

Based on this examination of the Canadian experience with networked operations, Babcock's draft definition of NEOps, given at the beginning of this book ("the conduct of military operations characterized by common intent, decentralized empowerment and shared information, enabled by appropriate culture, technology and practices"), appears to be the most suitable for Canadian purposes.[94] This is a much more human-centred definition of networked operations and is more compatible with Canadian military culture and experience than current American definitions. This recognition of the importance of the human dimensions of NEOps is reflected in the latest draft CF "concept and roadmap paper."[95]

6

Paradigms in Conflict: Critiques of Network-Centric Warfare

One of the major critiques of NCW is that it is a technology-centred approach to war fighting and other operations. Clearly, technology has had an important impact on how recent campaigns have been conducted. For example, in the post-September 11 epilogue to her book, Sloan argues that the Afghanistan campaign left almost no area of the RMA untouched, especially the use of precision munitions and disengaged combat. She notes that 60 per cent of the munitions dropped on Afghanistan were precision-guided, compared with 35 per cent for the Kosovo campaign and 6 per cent for the Gulf War. Furthermore, the first use of unmanned combat vehicles on a large scale has led to the prediction that by 2025, 90 per cent of combat aircraft will be unmanned.[1] Yet despite all these technological advances, some parts of the campaign were not much different than those waged eighty-five years ago on the Western Front. A recent "lessons learned" brief from Afghanistan pointed out that, like their First World War ancestors, US (and Canadian) ground troops were still lugging into combat eighty pounds of equipment on their backs.[2] This is only one example of how technology has not changed every aspect of warfare and why it should not be the focal point for future approaches to war and other operations. Some of the most trenchant criticisms of a technocentric approach to war and other operations follow.

US AND OTHER CRITIQUES

TECHNOLOGY AND HUMAN FACTORS A fundamental policy and budget issue for many armed forces today is what balance to strike between technology and human resources in force structures of the future. Often the question is framed as: what proportion of expenditures should be allocated to new equipment versus training? Stephen Biddle's iconoclastic interpretation of coalition success in the 1991 Gulf War offers a model that incorporates both factors. He uses it to support his premise that "future warfare is an incremental extension of a century-long pattern of growth in the importance of skill differentials between combatants," and that outcomes between highly skilled opponents have changed relatively little in spite of major changes in technology. His explanation of coalition victory in the Gulf War posits a powerful synergistic interaction between a major skill imbalance and new technology to account for its outcome. He theorizes that it was only the extremely low level of skill of Iraqi forces compared with that of Western forces in the coalition plus the technical preponderance of the coalition that allowed it to win a near bloodless victory. Biddle claims that higher Iraqi skill levels, even with their technological inferiority, would have resulted in significant coalition casualties; likewise, lower coalition skill levels, even with technological superiority, would also have resulted in significant coalition casualties.

Biddle maintains that his interpretation has important policy implications, because most current net assessment and force-planning methodologies focus on numbers and the technical characteristics of adversaries' weapons. These methodologies run the risk of producing a serious misjudgment of the real military power of opponents and could result in major errors in estimates in the forces needed to meet future threats. Biddle claims that those who argue that modernization should be protected at the expense of training and readiness overestimate the value of technology and underestimate the effects of skill in using technology on the outcome of a conflict. He concludes that a more systematic study of opponents' skills is needed because little research has been done on the relation between weapons effects and the skills of the operators.[3] Biddle's ideas have important implications for Canada and other medium powers, since potential US coalition partners must consider the trade-off between numbers and quality of troops and quantities of sophisticated equipment.

Owens cautions against putting technology ahead of other considerations, a phenomenon he labels as "technophilia." He argues that "Technophiles contend that a 'revolution in military affairs' based on emerging technologies has so completely changed the nature of warfare that many of the old verities no longer hold true. The technophiles argue that the US must do what is necessary to ensure its dominance in military technology even if it means accepting a substantially reduced force structure." But, Owens says, the future is unknowable, and the US has confronted at least one strategic surprise per decade since Pearl Harbor. He recommends not relying too heavily on technology and maintaining balanced forces that work together like the blades on scissors.[4]

Others suggest that the very nature of technology has changed at the beginning of the twenty-first century. Leonhard asserts that future war will be characterized by prototypes rather than mass production. Because of the rapid evolution of technology, he argues that there will be no "technological end state," but that in an era of technological flux it will be the side that can adapt and field workable prototypes based on changing permutations and combinations of technology that will succeed. This will be a major challenge to the American war-fighting culture, long based on quantity as much as quality, because the new "prototype warfare" will require "unprecedented levels of innovation and flexibility among warfighters."[5]

c² CRITIQUES Another major theme in critiques of NCW is that its implementation will have adverse and unintended consequences on command and control. The potential of NCW is huge if commanders were to actually have access to all the information that could affect their missions. With the holy grail of "full situational awareness" potentially so close to hand, many advocates see the developing technology as a panacea, without recognizing the extent to which it challenges traditional notions of command and control. Although primarily an issue that will be settled by the US armed services, the implications for America's coalition partners are huge. The Australian Defence Forces are in many ways comparable to the Canadian Forces. The following discussion from the Royal Australian Navy's Sea Power Centre, although made in the narrower context of undersea warfare (USW), neatly summarizes the scope of the issue:

[A] variety of [network-enabled] technologies promise to advance the sophistication of USW, offering the hope that increased mission effectiveness will derive from a combination of improved sensors, multiple platforms, and efficient, rapid data exchange and fusion. But there are profound difficulties in the practical application of both the technology and the doctrine. The larger debate about the nature and value of NCW is far from settled, and the debate about how to apply and manage it in the underwater battlespace is even less mature. ADF [Australian Defence Forces] doctrine acknowledges the as yet unformed nature of NCW and the risks inherent in trying to incorporate it into Australia's future warfighting concepts. What is clear is that we have not yet witnessed the genesis of either a concept or a technology that will make the oceans transparent. It also seems likely that rather than a revolution, NCW operations will ultimately be seen as another step in the leap-frogging process USW has followed since World War I. Certainly, there is nothing to suggest that the next two decades will witness other than a continuation of this process.[6]

Others are learning the lesson that allied navies have come to appreciate: the problem with NCW seems to be one of learning to filter the flow of all that information so that it reaches commanders at a manageable rate. It is easy to perceive many of these problems as limitations of current technology, but it is important also to acknowledge that there is a limit to what the human mind can process. There is a large and growing literature on some of the problems related to the human dimensions of command and NCW, and some important ones are summarized here.

Thomas Barnett, formerly a professor and senior decision researcher at the US Naval War College, offers a number of criticisms of NCW, but he is particularly critical of the strain the common operating picture could put on commanders at all levels. It may push too many commanders, fed by an almost unlimited data flow, into being control freaks, turning the common operating picture into a sort of non-stop internal spin control by commanders trying to influence what others see. It also risks becoming a command-manipulated virtual reality, at worst degenerating into the senior command staff engaging in a heavy-handed enforcement of the commander's view of the situation, all in the name of shaping and protecting the common operating picture. In any event, the developers of NCW may have fallen into the technology trap of providing information for information's sake, without considering the real needs of commanders.[7]

William Lescher, who reminds us that in large organizations the pace of innovation is constrained more by organizational culture than by technology, offers another caution. He argues that unless the US military gets past its fascination with technology to address critical issues such as a zero-defects mentality, risk aversion, poorly designed war-fighting experiments, and widespread contentment with current performance, expectations for NCW will not be realized.[8]

More recent criticism of NCW has addressed its conceptual origins. Kagan argues that the underlying flaw in NCW is that it reflects an effort to translate a business concept of the 1990s into military practice. The basis of NCW is drawn explicitly from the examples of companies like Cisco Systems, Charles Schwab, Amazon.com, American Airlines, and Dell Computers, among others. It has been claimed that all of these companies attained dramatic competitive advantages in their fields by creating vast and complex information networks and that using these networks to predict inventory needed to meet customer orders has permitted them to become "maximally adaptable," building products to the exact specifications of each customer only when the customer wanted them. This information technology allowed these companies to make enormous efficiencies because they could make accurate predictions, minimize risk, and adapt to rapidly changing circumstances. The key to NCW, according to its advocates, is to achieve information dominance over the enemy in much the same way that successful corporations use information to dominate their markets.[9] However, recent experience as well as history should remind us that war is not business, enemies are not customers to be serviced, and the type of information dominance this approach to war demands is unlikely to be achieved with enemies who are adaptable and able to foil attempts to gather intelligence, especially using the technical means that predominate in NCW.

Another problem with NCW, according to Kolenda, is the assumption that fusing information into a common operating picture will result in uniform interpretation of the information by its various users. He argues that shared situational awareness does not inevitably lead to "shared appreciation on how to act on the information," since different people, on the basis of experience, education, culture, and personality, will assess risk and how to best "maximize the effectiveness of themselves and their organizations" differently.

Kolenda concludes that technology-based common operating pictures can be used to keep the creative abilities of subordinates within the framework of a commander's intent; however, these subordinates must be given the authority and autonomy they need to create, within the commander's intent, original solutions to the problems at hand. Therefore, to ensure success, information technology should "result in empowerment and initiative rather than rigidity and overmanagement."[10] One tool to ensure the effectiveness of subordinates in networked operations is Pigeau and McCann's Balanced Command Envelope, as described in an NCW environment by Forgues below.

The most recent NCW policy statement, *The Implementation of Network-Centric Warfare*, offers evidence from operations in Afghanistan and Iraq of the success of NCW in these campaigns. But a number of commentators have observed that there has not been enough open debate on these lessons and that some of the lessons have been used to achieve conclusions that are personalized or politicized. As one commentator put it, NCW "has a 'certain naive quality' when it focuses on concepts like 'information dominance' to the exclusion of other ideas, including those that might undercut its value."[11] For example, immediately after what has been described as "the combat phase" of Operation Iraqi Freedom, many proponents of NCW declared that NCW had been responsible for achieving victory in Iraq. It soon became clear that there was much to be done before victory could really be declared. One American commentator reflected the views of a number of critics of NCW with this statement: "The Pentagon's version of 'transformation' is all about using technology to enhance the military's standoff power – the precision-guided bombs and unmanned robots that allow America to dominate a battlefield without risking high US casualties. But political transformation requires the opposite – an intimate 'stand-in' connection with the culture and people you propose to transform."[12]

Despite the optimistic quotes found in *The Implementation of Network-Centric Warfare* about the success of NCW in recent operations in Iraq and Afghanistan, reports from the field call them into question and highlight the Canadian perception that the human dimension of networked operations, like HUMINT (human intelligence), are more important in some circumstances than information gathered by technical means: "HUMINT drives successful operations and allows us to focus combat power, but we are happy if we hit pay dirt 25 per cent of the time. There is little useful information that comes down from higher, and the higher it comes from the less useful it is for a maneuver battalion. There are too many variables for it to be precise."[13]

Finally, others have criticized lessons from operations in Afghanistan and Iraq cited by advocates of NCW because the lessons were gleaned from "fighting incompetent adversaries" and were not "necessarily a good basis for making long-term force posture decisions" because future enemies may be more capable.[14]

One of the most controversial topics in the command and control and the networked operations debate today is the relationship between technology and the exercise of command. As Moltke observed over 100 years ago, it is important to recognize the limitations of technology and take them into account when exercising command. Van Creveld has expanded on this idea and, while noting that there is no such thing as a "one size fits all" c^2 system, has reminded us that there will always be unpredictability caused by the fog of war.[15] But as Robert Polk noted, this point is often lost on the "C4ISR" crowd, who believe that technology can tame uncertainty and that the future of war lies more in the art of mastering the science of well-laid plans than in fighting an opponent.[16]

This excerpt from a recent analysis of c^2 in an NDHQ staff paper summarizes the issues nicely:

"We want our leaders and their subordinates to be enabled by appropriate information technologies and architecture in order to develop the situational awareness essential for mission success. However, confident battlespace awareness will only result from the appropriate fusion of technology, organization, doctrine and personnel. There is no point in generating more information about the battlespace if: a) the doctrine is not well enough developed to assist in managing the information; b) the technology cannot rapidly and securely transfer vast amounts of data over long distances; c) the organization is so layered and compartmentalized that the right information never reaches the right people in time; and d) operators are unable to derive action-relevant knowledge from the information displayed to them."[17]

Another factor that should be taken into account in the command and control and the networked operations debate is the different physical, not to mention cultural, environments in which armies, navies, and air forces operate. Air forces operate in a relatively simple environment, and in this comparatively uncluttered battlespace command-by-direction and command-by-plan are not only possible but, as we have seen, perhaps necessary. Armies have, arguably, the most complex operating environment, and the stated command doctrine of most Western armies, mission command or command-by-influence, is designed to take this complex operating environment into account. Navies are, perhaps, in between the other two services and therefore susceptible to a greater range of command styles. While command-by-direction has been practised by navies for centuries, the last two decades have witnessed revolutionary progress towards command-by-influence in naval operations.

Given these factors, a useful starting point in the design of any new c^2 framework, but one rarely considered by technophiles who design systems to maximize technical possibilities (like bandwidth and resolution), would be an explicit statement of what the commander requires from the system. These requirements will of course vary according to the commander's role and, it would seem, according to a commander's personal qualities and preferences. This puts new meaning into van Creveld's observation that there is no such thing as a "one size fits all" c^2 system.[18]

CANADIAN CONCERNS

Many Canadian concerns about NCW as a basis for NEOps are similar to those summarized above; therefore, they will not be repeated in detail here. However, a number of concerns raised by Canadian commentators that provide a different dimension to the criticisms of NCW are worth discussing.

A c^2 CRITIQUE One of the earliest Canadian critiques of NCW was made by a student on the Advanced Military Studies course at the Canadian Forces College in 2000 and published in 2001 using the Pigeau-McCann framework as an analytical tool. At the time, many of the human-centred issues now appearing in NCW documentation were not found in the NCW literature.

Forgues notes that recent advances in information technology have affected the organizing principles for the conduct of operations and that NCW is one approach to "further exploit information technology and significantly enhance the functions of command and control on the battlefield." He argues that "command is a mission-oriented human endeavour performed within the limits of a commander's personal attributes," and that this requires "creativity and intuition to make sound decisions in a NCW environment."[19] One could also add that given the highly stressful nature of modern warfare, emotional and interpersonal competencies will be equally important for future leaders.

In the NCW environment, Forgues argues, the factor of personal authority "will create a double-edged sword that commanders will need to wield carefully." If things are going well during an operation in a networked environment, word of success will quickly spread, and this should increase a commander's personal authority within his force. But if the force encounters setbacks or failures, the commander's personal authority could decrease. Along these lines, another phenomenon not discussed by Forgues but that could be magnified by NCW is that commanders' decision processes may be more visible to subordinates at all levels, with positive or negative outcomes.

Another double-edged sword in an NCW environment may be the factor of intrinsic responsibility. The high degree of shared awareness that NCW should bring among members of a force could act to increase intrinsic responsibility in certain circumstances. For example, a setback might cause some elements of a force to have an increased sense of their responsibility to carry out a mission, but a shared awareness of imminent defeat might adversely affect intrinsic responsibility and paralyze the force. As we have seen, NCW may complicate or obfuscate lines of extrinsic responsibility.[20]

A networked environment might have a variety of effects on what Pigeau and McCann called "shared intent" among commanders and their subordinates. There is every reason to expect that explicit shared intent will be dramatically increased with NCW; however, Pigeau and McCann tell us that it is only the tip of the iceberg. The effect of NCW on implicit intent is not known at the moment, and yet it is arguably the most important of the two parts of shared intent. In any event, as Forgues observes, "the fundamental need of shared intent and the element of trust it engenders will remain a cornerstone of command in network-centric warfare."

In summary, based on the Pigeau-McCann framework, Forgues tells us that a commander should be within the Balanced Command

Envelope; in other words, "a given commander's abilities must match the levels of competency, authority, and responsibility associated with his position." Furthermore, he asserts that "information technology alone is not sufficient to enable self-synchronization in a NCW environment," but that organizations will "need to ensure that commanders at all levels have the attributes necessary to accomplish the task." Forgues concludes:

> The NCW environment will not determine the essence of command in war. The technology will indeed bring a new set of variables to the command equation that must be solved by commanders. In the words of Martin van Creveld, "Far from determining the essence of command, then, communication and information processing technology merely constitutes one part of the general environment in which command operates." The technological component of war can never fully account for the dynamic interaction of human beings and "war will remain predominantly an art, infused with human will, creativity, and judgement."[21]

In response to critics like Forgues, the most recent statements on NCW by Cebrowski have attempted to make the human dimension of NCW more prominent.[22] However, because of its theoretical and experiential roots, many see NCW as still excessively focused on technology. This is consistent with the view of those at the forefront of developing the NEOps concept in Canada who noted that NCW tends "to focus attention excessively on the network and its related technology, and [seems] to exclude military operations other than war."[23]

RECENT CRITICISMS In March 2005 two workshops were convened under the auspices of DRDC – Toronto to gather the views of Canadian subject matter experts (SMEs) on networked operations. The following excerpts from the resulting report highlight the main concerns of Canadian SMEs expressed at those workshops:

> Speed of command allows participants to adjust and modify their position more quickly, thereby leading to more robust Commander's Intent. A hidden assumption to speed of command, therefore, is that the locus of command can rapidly shift, i.e., "command is allowed to fluctuate" based on who has the

most relevant knowledge for the given situation. And this knowledge can be more than mere core knowledge. In response to this interpretation, however, one SME noted that this notion is possible in the Army, whereas it is difficult in the Air Force and Navy. For example, following 9/11, within the Air Force, the decision to shoot down a passenger airline emerged. As one SME explained, in an NEOps paradigm, a decision such as this would necessarily remain in the hands of the commander, because he is ultimately responsible for all activities and some decisions are simply "too important." In particular, there was a concern among SMEs that speed of command would lead to a faster means to make old mistakes.[24]

The ultimate outcome of NEOps is *increased mission effectiveness* [emphasis in original], which can be understood as quicker submission of the enemy with decreased lethality and destruction. Of course, within a peacekeeping operation, this would need to be defined differently. For example, one SME noted that mission effectiveness might be understood as "improving quality of life." Thus, the political and social outcomes are as important as military outcomes.[25]

Another key challenge in NEOps is the often implicit assumption that simply providing people with access to the same information will enable common understanding. Again, the issue of how "common intent" can actually be promoted among network players, often from diverse backgrounds and cultures (both national and organizational) represents a major challenge for the future. As such, there will need to be consideration around control mechanisms. For example, what is the role of doctrine and mission command?[26]

SMEs also noted that another potential challenge to NEOps will be attempting to implement it universally within the CF. In other words, SMEs argued that a "one size fits all" approach would undermine the particular nuances across environments in the CF. One SME believed that the impact will be more dramatic on the Army than the Navy or Air Force, explaining that the interaction of the soldier on the ground with another member of the land force is very different from the interactions in a maritime or air context. Some of the literature tends to support this perspective. For example, the notion of joint interoperability has been questioned because of the belief that air, sea, and land combine to achieve a "'unified' battlespace" (McMaster, 2003). But as McMaster states, "the factors that preserve uncertainty in war despite technological superiority are mainly land-based."

Finally, NEOps will be a challenge to the organizational culture and structure. According to MacNulty (cited in Warne et al., 2004), some changes to organizational culture will be reflected in command plans, the planning process, competition, attitude to change and risk, decision making planning cycle, and resourcing systems. Currently, there appears a lack of scientific investigation regarding NEOps and its impact on and interaction with CF culture.[27]

It appears, then, that the kind of transformation required for NEOps – or more specifically something like self-synchronization – will be a product of culture and doctrinal change within the CF as opposed to technological implementation.[28]

SMEs also noted that within the NEOps paradigm, the hierarchical structure of the military will be changed into a flatter organization, which resembles a "web of command" instead of a chain of command. If one of the desired outcomes of NEOps is distributed decision making, then the CF needs to consider the changes to the organizational structure that are required. For example, current C^2 is based on a central, hierarchical model. While thinking around greater horizontal command approaches has been emphasized (McCann & Pigeau, 2000), how does NEOps make this process more of a reality and hence more immediate? How does CF culture begin to embrace a "web of command" in place of a chain of command? This may require another form of leadership to reflect decentralized decision making, while still maintaining the essential level of authority. This leads to the question of how authority changes in a NEOps environment.[29]

SMEs also noted other challenges likely in implementing NEOps in Canada. Working within a JIMP context, for example, was seen as likely to present unique challenges to working in networked operations. For example, SMEs pointed out that although NEOps needs to be understood within a broader operational context, evolving partnerships (e.g. with differing JIMP stakeholders) will require different sharing requirements.[30]

Interestingly, the CF *Strategic Operating Concept* (2004) identifies the implausibility of removing all of the fog and friction of war through networks. It is documented that "human intelligence, obtained in part through human networking, will be key to achieving [an] information advantage" in the future battlespace. Though networks and sensor capabilities have improved the operational picture and decreased the uncertainty of war, certainty will never be realized because "[d]ifferences in individual cognitive processes, technological failures, and the actions of adaptive adversaries will all continue to frustrate achievement of a completely certain operating picture" (CF *Strategic Operating Concept,* 2004, p. 18). So despite the information advantage that arises from robust networking, commanders will still have to make decisions in the face of uncertainty. Networks themselves will not eliminate the uncertainty of war. These points highlight the caution in the Canadian perspective of NEOps when compared to the U.S. conception of NCW.[31]

However, there was a general concern among SMEs that, as the CF moves forward, it should not get "blind-sided" by the mere technological potential for combat operations. Rather, the CF also needs to embrace the full extent of transformation and the paradigm shift in military affairs and take into account the unique roles that Canada plays in international affairs. It also needs to consider the unique impacts that NEOps will have on the human actors and the CF organizational structure and culture. As such, SMEs identified a number of cognitive and social factors that require investigation as Canada moves forward. They feared that there might be many rapid organizational changes without the benefit of the robust research that they thought necessary. SMEs also thought that it was critical to integrate Canadian strategic operating concepts, such as the JIMP framework and the 3D approach, to international affairs through a fully articulated definition of NEOps. In fact, it was pointed out that NEOps is a governmental concept rather than a military concept. The question remains whether a military model will dominate in the governmental model. SMEs also thought it was important to differentiate the Canadian concept of NEOps from the US concept of NCW in order to ensure that all of the missions in which the CF participates are given adequate attention.[32]

Writing at the turn of the twenty-first century, Sloan concluded that an RMA was underway and that it has the potential to dramatically change warfare in the next two to three decades. Her conclusions about how Canada and countries with similar defence forces could deal with rapid technological change as represented by the RMA are equally applicable to NCW. She argued that despite the challenges of expensive equipment and small budgets, these countries can, by making selective investments in new technology, maintain some capabilities that will allow them to be interoperable with or provide niche capabilities to American and other coalition forces. She suggested that Canada invest in capabilities that can respond to both high- and low-intensity tasks, for example, advanced C4I, intelligence, surveillance, and reconnaissance systems, UAVs, strategic lift, precision-guided munitions, and highly lethal yet rapidly deployable and mobile ground forces. Sloan echoes Biddle's concerns when she advises that Canada must consider the trade-off between personnel and technology. She concludes that to ensure this trade-off is set above the line of operational and political marginalization, increased defence spending is required.[33] It remains to be seen whether recently promised increases in Canada's defence budget will be enough to address her concerns.

7

Other Approaches to Networked Operations

There are a number of ways of approaching the transformation of armed forces to deal with future war and other operations, as we have seen. In the previous three chapters some of these approaches, based on different services' paradigms of war fighting and the Canadian experience of operations, were examined. In this chapter, two other approaches are discussed. The first, Fourth Generation Warfare, has assumed a place of prominence lately because of recent operations in Iraq and Afghanistan. The second approach, which is based on recent Canadian Forces experiences in post-Cold War operations, is evolving.

FOURTH GENERATION WARFARE

A leading challenge to the notion of the technically oriented information-age warfare espoused by NCW advocates as a way of viewing future warfare is the more people-oriented concept of Fourth Generation Warfare (often abbreviated as 4GW). In 1989 William Lind and four co-authors coined the term, and despite its shortcomings as a model as identified by some,[1] it has gained some currency in debates about the future of war. At that time Lind and his colleagues identified three historical generations of warfare and two possible future fourth generations. They categorized the first three generations of war as follows: First Generation War (1648–1865) was fought by state armies using line and column tactics; Second Generation Warfare relied on firepower to cause attrition and was described as "war by body count"; and Third Generation Warfare was a German product, fought more in time than in place, and based on speed and manoeuvre.[2] They characterize the American way of war, even today, as Second Generation Warfare because its goal is still "victory through attrition" and because the new technology (like the B-2 Stealth bomber and the Predator UAV [unmanned aerial vehicle]) in the current US transformation strategy is only designed to make firepower more efficient or more "precise."[3]

In the 1989 article the authors provided two possible models for 4GW: technology-driven warfare and idea-driven future warfare. In technology-driven warfare, technologies like directed energy weapons and robotics will allow small, highly mobile forces combined with information operations to attack an enemy's centre of gravity, which is defined as the enemy population's support of its government and the war. This type of 4GW posits a state-versus-state conflict, but recent events have caused Lind with his co-authors and other writers to focus on the second hypothesis for 4GW.

Noting that the events in Iraq from 2003 to the present have marked "the end of the state's monopoly on war,"[4] Lind and others have turned their attention to idea-driven 4GW, which according to Wilson et al was cited in an Al-Qaeda-affiliated Internet magazine as the foundation of Al-Qaeda's military doctrine.[5] In the 1989 article, Lind et al asserted that even though terrorism is neither new nor particularly effective, if combined with new technology it could be extremely potent. A number of others were also aware, pre-9/11, in general terms, of possible terrorist threats. For example, in the early 1990s van Creveld observed that since 1945 most wars had been fought by small, concealed, dispersed groups of terrorist organizations that had no clear territorial base and could not be targeted by modern technology.[6] Others note that the concept of "small wars," or guerrilla warfare, have ancient origins and modern roots in the Napoleonic Wars, where the term guerrilla (the Spanish diminutive of *guerra* [war]) was coined.[7] Many observers now agree that idea-driven warfare is commonplace around the world and that 4GW foes can attack the entire social order by using the target society's own organization, laws, technology, conventional forces, and tactics against itself. Opponents are therefore using 4GW concepts to leverage the Western dependence on technology and to avoid a decisive fight, using "4GW judo" to keep large Western security, military, and legal bureaucracies off balance.

Yet how the West can deal with foes using 4GW concepts, where the tactics of the weak confound the tactics of the strong, is still not well understood. A number of commentators argue that Western "targeteers" (using 2GW) are defining and attacking artificial, physical enemy centres of gravity with precision weapons – bringing to mind an old adage that when your only tool is a hammer, all of your problems look like nails – whereas the real centre of gravity is a shared religious or ideological goal where common purpose and zealotry replace military equipment and command structure. Wilson, one of Lind's co-authors of the 1989 4GW article, concludes that "as technophiles Westerners are enraptured by weapons of great precision but have lost sight of the fact that people and ideas are the essence of why wars are fought and for how long."[8] Therefore, advocates of a technical revolution in warfare may be using a dated twentieth-century paradigm to interpret change in war, when the problems of linking technology and doctrine are much the same as they have always been. Some of those who challenge NCW as a concept believe that because its focus is on technical solutions to problems encountered in twenty-first-century operations, it does not adequately take into account the human dimensions of implementing these solutions. This leads to the idea that the human network, not the technical network, might be the basis for future theories of war.

EMERGING CANADIAN CONCEPTS: CREATING AND USING HUMAN NETWORKS IN AFGHANISTAN[9]

In 1961 J.F.C. Fuller wrote, "the true aim of war is peace and not victory; therefore that peace should be the ruling idea of policy, and victory only the means toward its achievement."[10] Although he used these words to discuss the legacy of nineteenth- and twentieth-century warfare, they seem prescient at the dawn of the twenty-first century. It has once again become apparent, most recently in Afghanistan and Iraq, that in addition to combat operations, successful military campaigns must also emphasize stability and humanitarian efforts to ensure a long-lasting peace. During Operation Iraqi Freedom, Commander V (US) Corps, Lieutenant General William Wallace, was quoted as saying, "One day our troops are kicking down doors, and the next they're passing out Band-Aids. And in some cases, they're kicking down doors without really knowing if they are going to have to pull a trigger or pass out a Band-Aid on the other side."[11] The dilemma of operating in an environment that has come to be known as the "three-block war" provokes a great deal of thought on the nature of crafting methodologies that will encourage peaceful societal rejuvenation, as a collective effort by a variety of organizations.[12] In this context successful military intervention requires not only the use of decisive force and an implicit understanding of the nature of reconstruction and renewal required for a particular region but, very importantly, an ability to establish and maintain human-centric networks. Members of the CF are continually thrust into environments that require the use of these sorts of concepts, and its members continually rise to that challenge and establish human-centric operations. To gain insights into how men and women of the CF create effective military intervention in post-conflict situations, one can examine the work conducted by the Canadian-led International Security Assistance Force (ISAF) Rotation V in Afghanistan during 2004.[13]

The situation in Afghanistan at the time of the first large-scale Canadian deployment with ISAF Rotation IV in 2003 was stable, but fragile. Afghanistan had endured decades of strife and friction throughout its borders that had destroyed any confidence in the ability of central authorities to be able to address issues of governance. At that time an American officer described the situation unambiguously:

> There are clearly things that need doing here – from basic services to CCC [Civilian Conservation Corps]-like projects (roads, reforestation, irrigation, etc.), to building of every governmental institution ... We need to disarm ... and teach them about other ways of life besides poppies and illegal ... check points ... from which they make a living and fund and pay their armies ... But it's not about physical building (although that's part of it), it is about building the ability for Afghans to take care of themselves from public health to judicial systems. These are much harder to grow than trees or roads or schools ...[14]

This dilemma was highlighted by Canadian Prime Minister Lester Pearson in his 1957 Nobel prize lecture: "Our problem, then, so easy to state, so hard to solve, is how to bring about a creative peace and security which will have a strong foundation."[15] To address these challenges it is essential to produce societal renewal by understanding the nature of the culture being rebuilt and by working within that society.[16] This requires focusing on the processes that permit strengthening and development of internal structures. These processes are predicated not on technical connectivity but on the establishment and maintenance of people-to-people contacts. In this manner one can endeavour to ensure balanced efforts by all agencies to create the circumstances necessary for success. The roles played by internal and external participants in the process bear particular scrutiny because it is necessary to include local ownership in these networks as they evolve and mature. Military commanders and staffs must recognize that there is a need to subordinate the martial aspects of the intervention to the imperatives of supporting the efforts of reconstruction and provide the security that will encourage and enable these activities. While technology is of importance in establishing connectivity, in an environment that does not have a homogeneous technical network available to all participants, the networks that are established are ad hoc and hybrid. It requires great effort to construct these hybrid networks in order to create cooperation and interaction with a myriad of other groups as well as a cohesive and focused plan that fosters unity of effort. Unfortunately, this unifying concept rarely exists as a cohesive whole in either national or alliance strategy. However, as a result of Canadian experiences of peace support operations since 1992, the CF (particularly the Army) has developed the institutional knowledge to address these complex dilemmas.[17]

Peace support operations are very important to Canada, and the current Canadian vision of conducting these operations has been shaped by the experiences of the twentieth century. The national perspective of peacekeeping was initially framed by the extremely successful Canadian organization and leadership of the United Nations Emergency Force (UNEF) I in 1956 and by succeeding "traditional" peacekeeping missions.[18] Military historian Desmond Morton alludes to the impact that the 1956 operation had on the Canadian psyche with the comment, "Many Canadians were pleased to be useful."[19] This desire to be relevant continues to shape Canadian public perceptions of peacekeeping in the modern era. All of these missions were predicated on obtaining consensus from the belligerent parties through mediation and negotiation. Avoidance of conflict was the rule, and the maintenance of peace, or at least an absence of overt violence, was normally sufficient to achieve the objective of the mission. National and alliance concerns were primarily the containment and de-escalation of fighting, and Canada, as a post-war middle power, had justifiable pride in this role within the UN and NATO.[20]

However, in the volatile international environment of the 1990s and beyond, it became obvious that this paradigm was no longer workable. The goal of intervention is no longer simply a cessation of violence to avoid potential nuclear conflagration but reconstruction, renewal, and development leading to functioning nation states that can act autonomously within the community of nations. Military operations must assist with creating the conditions for a durable and lasting peace in joint, multinational, and multi-agency environments, with numerous state and non-state actors involved in the crisis. The military contribution to this goal is not limited to the separation of consenting belligerents but has become exceedingly complex. This has resulted in acknowledgement of the necessity that the CF must establish connections with the Department of Foreign Affairs and the Canadian International Development Agency to address the modern dilemmas of post-conflict Afghanistan. Then Defence Minister David Pratt articulated this idea in an address to the Conference of Defence Associations Institute during 2004: "One of the things that distinguishes Afghanistan from previous Canadian Forces missions is the unprecedented cooperation we're seeing between the Canadian Forces, the Department of Foreign Affairs and the Canadian International Development Agency. In fact, from the standpoint of future Canadian international engagements, Afghanistan is serving as a model for the government's 3D approach to international affairs – the three Ds being defence, diplomacy and development."[21]

The 3D approach has required an increased level of interoperability between departments that lack a common information infrastructure. The result has been dependence on social networks and the establishment and coordination of decentralized operations. The sensors, processors, and systems used to create lethal and non-lethal effects in such an environment mirror many of the principles of NCW.

This phenomenon is similar to Canadian experiences in overseas operations, such as the later phases of Canada's mission to Bosnia during Operation Palladium, and domestic operations, such as those conducted during the Winnipeg flood or the Year 2000 (Y2K) contingency planning. The creation of these hybrid networks is normally predicated on the support of military organizations to multi-agency efforts in complex environments. Military headquarters are trained, structured, and resourced to provide the necessary functions that will encourage the establishment, maintenance, and coordination of all activity. Intertwined with these aspects is the realization that efforts to ensure a solidly constructed peace in Afghanistan require a great deal of perseverance and patience. This determination is extremely difficult to sustain over great lengths of time, particularly with the improvised nature of the networks that are established. As a result, military commanders and planners must be prepared to provide moral and physical support to the other non-military agencies involved in the effort. Incidents that have occurred in other peace support operations, such as the riots at Drvar, Bosnia, in April 1998, have posed great challenges to international and non-governmental organizations and may result in a temporary or even permanent cessation of activities unless the necessary partnerships are in place.[22]

This provokes a great deal of thought on the exact nature of military assistance and its role within peaceful societal rejuvenation as a collective effort by a variety of organizations in the context of an overarching strategic concept. It is obvious that the successful application of military and other aid requires not only an implicit understanding of the nature of societal reconstruction and renewal needed for a particular region but also linkage to multi-agency structures that provide focus to all entities within a given country, like Afghanistan. In the absence of this arrangement it may then become necessary for the military to provide the assistance and impetus necessary for the formulation of an overarching system, or networked community, with an ultimate objective of resolving "the removal of the causes of war."[23] This was the approach used during ISAF V and was the product of Canadian experiences with the complex dilemmas posed by peace support operations since 1992.

Before the transition between ISAF IV and ISAF V in January 2004 the situation in Kabul had been stabilized through intensive patrolling.[24] The time was propitious to expand beyond tactical activities and implement a long-term plan. In a meeting preceding the NATO changeover, President Hamid Karzai, the leader of the Afghanistan Transitional Authority (ATA),[25] expressed concerns regarding four principal areas that he felt would undermine the influence of the ATA. First, there were internal and external threats to the fledgling

Afghanistan National Army (ANA) from a variety of sources. Second, the lack of human capital, that is, educated and trained people, within the ministries of the government and the security forces of the ATA posed great challenges to the governance and security of the country. Third, there was a need for the promulgation of positive information supporting the efforts of the ATA. Last, and most important, the absence of unified action by the multitude of governments and organizations in Afghanistan had resulted in a dissipation of development efforts and effects. As a consequence, to expand ISAF V beyond tactical and short-term activities it was evident that it would be necessary to deal with these concerns. The most important area to resolve was the lack of unity of effort amongst those involved in the regeneration of Afghanistan.

On account of this apprehension and as a result of Canadian experiences in the former Yugoslavia, it was recognized that the primary focus of ISAF V ought to be the development of a strategic objective supporting the harmonization of the international community's efforts in the reconstruction and redevelopment of Afghanistan. The initial construct was created by a Canadian-led planning team that reviewed the mandates of all of the major organizations in Afghanistan and compiled a list of the objectives of each one. This proposal was first known as the Structured Process for the Harmonized Development of Afghanistan (SPDA), and later the Investment Management Framework (IMF). Depicted in Figure 7.1, these independent but interrelated themes culminate in one national goal or "end state" and were represented within the IMF as five strategic lines of operation, each with a clearly defined start point and final objective. Each line of operation (Security, Islamic Republic Governance, Rule of Law, Building Social and Human Capital, National Economy and Physical Infrastructure) contained multiple objectives: short, mid- and long term. The goal of each line of operation would be attained gradually through these three successive phases.[26] The IMF model was offered to President Karzai and the ATA to further assist in developing a common national vision and unity of effort through consensus building and coordination of all parties. It also provided a method of prioritizing the challenges facing the ATA, with the purpose of eventually building a legitimate and functioning state that provided for the security and prosperity of its citizens and contributed to regional and global stability.

The IMF and derivative products that outlined the tasks and inter-dependencies required to achieve the objectives of the five thrust lines were briefed to the ATA and various members of the international community, including the UN Assistance Mission in Afghanistan (UNAMA). There was general acceptance of the concepts and principles that underpinned the IMF. Next, members of the Canadian-guided ISAF V planning staff assisted the ATA with incorporating this information into the National Priority Programs (NPPS). With the assistance of ISAF staff, about 40 per cent of the IMF was actualized through the NPPS.

The NPPS were created by the ATA towards the end of the first two years of their administration to move reconstruction efforts from discrete and easy-to-implement projects intended to provide immediate relief to those that would contribute to a sound future vision for the people of Afghanistan. The overarching elements of this vision were first outlined by President Karzai at a number of international and domestic conferences, and he proposed that policies entailing large amounts of funding should be carried out through a coherent program of development as opposed to the financing of discrete projects. This idea became the National Development Framework, and its component parts became the NPPS. The unifying principle of the NPPS was that donor aid should be allocated through the national budget process so the capabilities of Afghan institutions could be consistently and systemically increased. The NPPS were designed to move Afghanistan from a position of recovery and reconstruction to one of sustainable development as well as to address a number of areas that promote prosperity and deal with poverty, as the foundation of Afghanistan's future. Additionally, the NPPS would ensure that aid efforts would be "transparent, effective and accountable" to the populace of Afghanistan.[27]

While a great deal was accomplished with the creation of the IMF and the incorporation of a sizable portion of its contents within the NPPS, it was recognized that there was a need for a nationally unified concept that clearly articulated a strategy for prioritization and subsequently focused development. The next logical stage was the creation of a national concept for structured regional development. This concept would link the resources of a strategy with the methods of the NPPS to actualize the goals of the IMF. The resultant national strategy would provide crucial guidance and impetus for

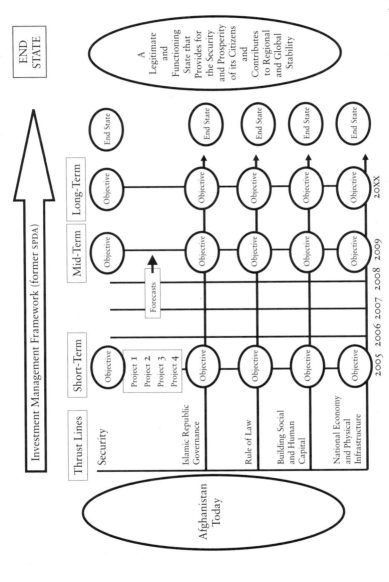

Figure 7.1 Structured Process for the Harmonized Development of Afghanistan (SPDA) and later the Investment Management Framework (IMF).
SOURCE: Hope, "A Strategic Concept for the Development of Afghanistan."

coordination that would facilitate the construction of operational-level planning for ISAF.

As a result of discussions with the ATA Ministry of Finance, the commander of ISAF V agreed to assist Finance Minister Dr Ashraf Ghani and his staff with the analysis of this strategic problem, and to augment ISAF V capabilities two officers deployed from Canada from June to August 2004 to fulfil this obligation. Their work and that of other members of ISAF V resulted in the concept paper "Creating a National Economy: The Path to Security and Stability in Afghanistan."[28] This proposal used the existing work of the ATA as a model for creating a legitimate and functioning Afghanistan. Without changing any of the established and accepted programs and policies, this concept paper advocated unified action of all involved agencies within an overarching security context: "The fundamental issue right now is security ... because there is clearly right now – given the security questions – reluctance [to make capital investments in the country] ... Where it may not be possible to secure the entire country at one point, you could create zones of security where enhanced economic activity could be fostered."[29]

The ideas contained in this concept paper were derived from an analysis of "obstacles to success" as well as various official documents, primarily the ATA paper "Securing Afghanistan's Future."[30] The greatest challenge to the re-establishment of Afghanistan was the lack of confidence in nation-building efforts by the international community and the Afghan populace. Four disintegrating influences contributed to this lack of confidence. The first was the political dissention of regional leaders who advocate local interests above that of the nation. The second was military and can be seen in the existence of the various disruptive non-governmental armed forces. The third were the people of Afghanistan themselves, who, having been fragmented by almost thirty years of violence, had little unity. The fourth were regionally based narco-economies that had global implications. These dissipating forces had come together in five distinct regional units that formed solid resistance to centralization under the ATA. Figure 7.2 approximately delineates these regions.[31]

The analysis contained in "Creating a National Economy" outlined a number of significant issues that needed to be addressed so that ATA and international efforts in Afghanistan might prove fruitful. Primarily, in the absence of a strategic coordinating mechanism

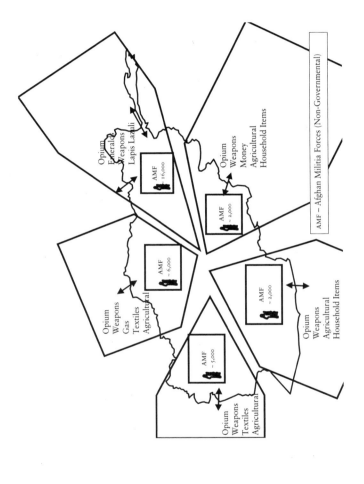

Figure 7.2 Obstacles to success in Afghanistan: religionalism.

SOURCE: Hope, "A Strategic Concept for the Development of Afghanistan."

and a common vision for development, the multitude of efforts by the ATA and international community were diffuse and consequently made little impact. Moreover, without the creation of a secure environment for those engaged in the re-establishment of the Afghan state, reconstruction efforts had stalled. It was also necessary to resolve the obstructive regionalism supported by the various narco-economies and orchestrated by various regional power-brokers.[32] Finally, there was a need to reinforce central government and the institutions of subnational administration so that they could provide effective governance and remove the perception that the ATA had little influence beyond the environs of Kabul.

"Creating a National Economy" also articulated a coherent idea to achieve a strategic vision of a rejuvenated and independent Afghanistan as an integrated member of the global community. It proposed that through focusing resources in specific areas, one can make the greatest impact in addressing these challenges, and it asserted that the ATA and the international community needed to combine efforts to transform regional illicit economies into one legitimate national economy. The best way to do this was to focus on the greatest vulnerability in each regional economy, specifically the rural areas where opium was produced. In cities and large garrison towns there was a concentration of political, military, and financial power that in most cases gave security and often prosperity to the urban populations, and acted as a reinforcing feedback mechanism to continued regionalism. In the rural areas there was no such concentration of military or financial power, and political power was also diffused. The most important groupings were those of village and family. One cannot underestimate the impact of these kinship networks, and their support would be necessary to enact this transformation. As the central government was challenged to extend its influence into remote areas, so were regional power-brokers. This weakness in the illicit economies permitted an avenue for intervention. Therefore, it was crucial to identify a limited number of regions – perhaps six to eight – in which to focus these efforts. These regions should be high-density poppy cultivation areas and have security situations that would facilitate intervention. The Afghanistan Opium Survey of 2003 suggested that the provinces of Ghor, Helmand, Baghlan, Badakhshan, Oruzgan, Kandahar, Sar-e Pol, Balkh, and Wardak contained such areas.[33] Further surveys of these provinces would provide the most suitable districts for intervention through concentrated development. A diagram of high-density poppy-growing areas is depicted in Figure 7.3.[34]

It was noted separately that one exception to this approach may have been the the city and province of Kabul. It was necessary that the ATA build credibility by addressing security issues in the capital region before attempting to do so elsewhere. Without demonstrating the ability and determination to create a safe environment in the immediate area of the capital, it would be that much harder to expand central governance elsewhere and enact the recommendations of "Creating a National Economy."[35]

This proposal for a focused program of development was a macro-model that allowed the ATA to take the lead in the re-establishment of national governance in Afghanistan. The analysis recommended using overarching planning and coordination mechanisms within the NPPs, with inclusive multi-agency representation to oversee the implementation of this idea. This coordinating body would then be responsive to the ATA and have subordinate coordinating committees in the districts being developed. In this manner a cohesive and unified approach, marshalling all resources, could be taken towards regional development. It was also acknowledged that this strategy could only work if there was consensus amongst all stakeholders created by leadership from the principal ambassadors, supported by the heads of donor organizations, in assisting the ATA to choose regions for concentrated development, in selecting priority institution-building programs, and in distinguishing work requirements.

The resultant efforts to achieve this outcome demonstrated the link between doctrine, NEOps, and EBO, in addition to the relevancy of current Canadian Army doctrine vis-à-vis the contemporary operational environment as articulated in this Army publication: "Network-enabled warfare will shift the traditional emphasis on platforms to focus on a system of highly integrated networks. Such a shift will allow for the application of the full range of non-lethal and lethal effects, including joint, interagency and multinational. These networks will fuse the available information using leading-edge technology to provide commanders with the best possible situational awareness so that specific effects can be brought to bear in a precise and discriminate manner."[36]

While best attempts are made to use technology in the manner described, more often than not the hybrid networks established are a mixture of secure and non-secure or Internet systems with human

SCOPE OF OPERATIONS:

• Select high-density poppy districts to receive intensive development packages.

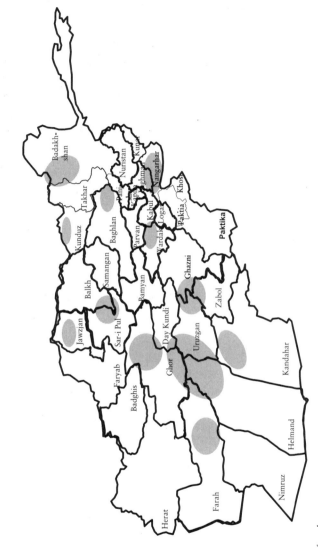

Figure 7.3 Concept for change.
SOURCE: Hope, "A Strategic Concept for the Development of Afghanistan."

interfaces, mechanical, and human sensors that create shared aware-ness with a common operating picture that is available to many enti-ties, both military and non-military.

In Afghanistan it was necessary to establish, amalgamate, and reinforce human and technical networks to create the desired effects. This indicates a key difference between visualizations of "normal" operations and peace support operations as they have come to be practised by the Canadian Forces. In "kinetic" types of operations the focus is on the destruction of the enemy rather than consolidation of one's own forces and authority. However, it is important to recognize that such approaches in the tumultuous environment of a war-torn nation may not always succeed because the problems are multi-faceted and not easily resolved. As a result it is necessary to move beyond military thought that evolved in response to the demands of industrial age warfare, in which nations fielded mass armies against one another as a primary means of mili-tary activity. When the complexity of the problem changes from kinetic operations to nation building, often in a non-permissive environment, it is necessary to establish hybrid networks that address the challenges of post-conflict environments, including that of distributed operations against a networked opponent.

As noted above, early theoretical descriptions of NCW focused on its technical aspects, but more recently the United States Depart-ment of Defense Office of Force Transformation suggested that the human component of NCW needed to be considered:

> NCW generates increased combat power by networking sensors, decision makers, and shooters to achieve shared awareness, increased speed of command, high tempo of operations, greater lethality, increased survivability, and a degree of self-synchroni-zation ... In essence, it translates information advantage into combat power by effectively linking friendly forces within the battlespace, providing a much improved shared awareness of the situation, enabling more rapid and effective decision making at all levels of military operations, and thereby allowing for increased speed of execution. This "network" is underpinned by information technology systems, but is exploited by the Soldiers, Sailors, Airmen, and Marines that use the network and, *at the same time, are part of it* [emphasis added].[37]

This is the primary difference between the development of NCW in the Canadian and American circumstances. In 1998 the original concept of NCW outlined by Cebrowski and Garstka focused on emergent technology and the potential of an RMA. For the Canadian Army, a force that has not had a similar access to developing information management tools, the focal point has been human-centric networks incorporating technology to meet the challenges of the post-Cold War era.

Consequently, during peace support operations, such as that being conducted in Afghanistan, one must constrain the growth of the threat forces and manage the perception that there is an increase in measurable government capacity through information operations conducted by hybrid networks. In this distributed system the military component is just one element of the overall campaign. In certain ways this may have become a de facto Canadian way of conducting war and winning the peace. While deployed, officers and non-commissioned officers shape the environment through a variety of non-lethal means. This is necessary since kinetic operations do not address all aspects of often multifaceted dilemmas in post-conflict environments. For example, in Afghanistan the narco-state has arrived and military-generated kinetic operations alone will not deal with the convoluted issues this predicament creates. To be successful in this environment, we need to develop and refine linkages between all actors in the joint, multinational, multi-agency, and indigenous environments. The problem identification and analysis conducted by ISAF V in Afghanistan and articulated in "Creating a National Economy" expressed a strategic objective but required the establishment of social and other networks to promulgate and actualize the strategy.

As a result of the requirement to create agreement for the ideas contained in "Creating a National Economy," it was necessary to encourage communication, feedback, and eventually accord amongst all potential participants, including the ATA. Accordingly, a small group of Canadian-led ISAF officers worked diligently at facilitating common understanding and acceptance of this analysis. In light of the planned presidential and lower-house elections and likely leadership changes during the fall of 2004, it was somewhat difficult to deduce who would be in a position to become the prime proponent of this idea. However, there were a number of officials then in positions within the ATA who could act as advocates for this concept in the foreseeable future, and these individuals were briefed on the contents of this proposal.

Simultaneously a number of the Western ambassadors and their staffs were approached and demonstrated varying degrees of interest. The Canadian ambassador was informed of the initiative, and a copy of the package was sent to the Department of Foreign Affairs and International Trade for their information and further promulgation. Various international and non-governmental organizations were also solicited for feedback, and all showed support for some if not all aspects of the concept. It was also proposed that another possible venue for dissemination would be to forward the paper to consulting companies currently involved in Afghanistan for their information and to take advantage of their contacts and information. Wherever possible the paper was distributed to consultants and contractors because they are now part of the contemporary environment of peace support operations and have access to a multitude of networks.[38]

Information packages and briefings were also given to Coalition Forces Command – Afghanistan (CFC-A) to ensure that they were aware of the work that was ongoing and to gain their support.[39] The coalition and NATO forces in the provinces that were coordinating reconstruction efforts and would eventually have to act in an enlarged role were the Provincial Reconstruction Teams. They were capable of making the linkages between the ATA officials and the international community to assist with coordination and capacity building. They could also act as the military representatives in a given area and speak on behalf of their commanders. In a similar manner as the joint commission observers of the NATO Stabilization Forces in Bosnia, they would also be a "directed telescope" or way of passing information outside the usual reporting channels.[40]

Although time did not permit before the transfer of authority from ISAF V to ISAF VI, it was thought that it would have been useful to approach the World Bank and the Asian Development Bank to solicit their opinions on a unified approach to the reconstruction of the state of Afghanistan. Numerous experiences with stability operations in the recent past have indicated the need for active involvement of donors to achieve strategic goals.

This effort to create networks that would promote interest and shared ownership amongst all agencies was crucial. While it took a great deal of work, it was recognized that without assisting the ATA to advance the concept, both the populace and the international community would be reluctant to sustain it. The support of the international community was necessary for unified action, and without the security, planning resources, and coordination provided by the military, the ATA and international community would be reluctant to commit resources to this strategic vision. The hybrid networks so established also provided the sensor capabilities necessary to establish a common understanding of activities transpiring in all regions of Afghanistan and bordering countries. In many ways it becomes the role of the military commander to create and sustain these networks.

It is necessary to approach the difficult quandaries posed by peace support operations in post-conflict situations holistically and identify points that must be addressed simultaneously, in a distributed fashion, across elements of national power in order to achieve the desired result. When the opponent is an ill-defined and usually a non-state actor, activities during the planning process should be focused not on traditional views of overcoming threat forces but instead on the attainment of the objective of the operation by linking the people and organizations necessary for success. Additionally, it is very important that all military actions are conducted in a manner that will bear public scrutiny. Indigenous participation, support, and, most importantly, leadership of the process of reconstruction must be encouraged. Any concepts designed to assist with the renewal of war-torn societies must recognize the need for local participation in the transformation as it evolves and matures.

Although peace should be the ultimate aim of war, it is unfortunately often not the result. Retired US Marine Corps General Arthur Zinni highlighted this shortfall during an address to the United States Naval Institute in 2003: "Whatever blood is poured onto the battlefield could be wasted if we don't follow it up with understanding what victory is."[41] To bring into being a truly successful post-conflict peace, it becomes essential to produce societal regeneration by understanding the nature of the society being rebuilt. This requires focusing on the processes that permit strengthening and development of structures that promote networked operations. Specific military plans must be formulated that include unified and balanced efforts by all agencies to achieve the conditions necessary for success. Ideas contained within the 3D construct will aid in this endeavour within an inter-agency milieu. The roles played by internal and external participants in the process bear particular scrutiny above and beyond the need to recognize local ownership of societal transformation as it evolves and matures. The degree of access to the larger networks must be delineated. Commanders and staffs must address the need to subordinate aspects of the military campaign to ensure the involvement and engagement of the international community in networked, distributed operations.

As a result of the quandaries posed by the current situation in Afghanistan, the priority of security and development efforts must be the transformation of regional illicit economies into a legitimate, national economy. This is consistent with the strategic vision for the country and, as an essential condition for progress towards peace, stability, and civil society, it is a stated or implied goal of many current development initiatives. The combined resources exist in Afghanistan that, when reinforced by the diplomatic efforts of the international community, will achieve the objective of economic transformation within specific regions. The most viable way to do this is to target specific high-density poppy cultivation districts for substantial development intervention and connect them to urban, market, transportation, and trade development and local security sector reform activity. Military activities in this environment need to be prepared to bring stability to unstable places by promoting focused policy and by being ready to support development activities with military efforts that embrace the notion of the three-block war, combining elements of combat, stabilization, and humanitarian operations.

In the post–Cold War era, members of the Canadian military are continually thrust into complex environments without clear strategic objectives save for the imperative to establish a "secure environment." To function efficiently at the intellectual plane encompassing the three-block war and 3D approaches, the CF must be aware of the challenges experienced in the recent historical past, analyze them, and institutionalize the results in the methodology used to operate in these complex and difficult situations. The most significant lesson of current peace support operations is that an intervention by outside forces in a war-torn environment does not necessarily lead to a strategically certain result or a complete cessation of hostilities between the nations or groups involved. A tenuous peace in Afghanistan has once again highlighted the difficulties inherent in the re-establishment of a nation and confirmed the importance of hybrid networks that link military and non-military activities while allowing focused and unified activity. Only by establishing that connection can one orchestrate the elements that are required to build an enduring peace in post-conflict environments. In many ways the ability of members of the CF to create these linkages has become a distinguishing mark of Canadian military operations at the dawn of a new century. These recent Canadian experiences are just now coalescing into a new doctrine that may be the basis for NEOps.

8

Conclusions

NEOps seems poised to become the driving concept behind CF trans-formation for a number of reasons, not the least of which is Can-ada's tendency to follow the American lead in new concepts related to war and other operations. This study examined NEOps and its progenitor, NCW, and concluded that Canada and the CF should be cautious about using NCW as the basis for NEOps, because the con-text and needs underpinning NCW may not be congruent with Cana-dian requirements.

While the notion of networked operations has been embedded in conceptual approaches to operations of a number of militaries, recently a specific variant, NCW, has come to dominate the debate on change and transformation, and it is being used as a template for future American command and control frameworks. This domi-nation came about not because of any overwhelming empirical evidence or because NCW has wide-ranging practical virtues but because it was imposed on the US Office of Force Transformation by one of its leading advocates, the late Arthur Cebrowski. As we have seen, however, there is still considerable confusion as to what the concept of NCW actually entails because it has been evolving over the past seven years and because of its arcane language. Fur-thermore, as the concept has evolved, it has moved well beyond its naval roots and incorporated a number of models from other domains, including EBO, information age warfare, mission com-mand (or command-by-influence), manoeuvre, and elements of the OODA loop, which are not necessarily compatible with the original NCW construct nor always well articulated. This has caused a great deal of confusion in the debates on NCW-driven transformation.

Unfortunately, this confusion has been glossed over in a number of official publications and is exacerbated by the fact that even "transformation" is not clearly defined by those in charge of these efforts in the US today.

This study began by asserting that to adapt to change through innovation, military professionals and those in the defence community need to understand the intellectual as well as the technical tools they use in their work. To gain an understanding of NEOps as a professional tool, they must be conscious of the historical and theoretical context in which it originated and is evolving. We noted here that NEOps is seen by many as a branch on the NCW theoretical tree. However, NCW is not a theory of war in the sense of an idea or principle that has explanatory or predictive value; rather, it has been described as a series of largely untested hypotheses or assumptions strung together in various official documents. Moreover, proponents of NCW, while recognizing the value of a wide range of theories of war, have used the Tofflers' "third wave" information-age model as the theoretical foundation for NCW. This model has been widely criticized for its over-reliance on technological explanations for changes in war, and this is one reason why NCW has been characterized as a technophile's approach to war. It would seem prudent, therefore, to base any new approach to future war on a synthesis of various theories of war that incorporates their best features, rather than on one controversial model.

Each nation and each service in a nation's armed forces has its own unique paradigm of how military operations should be conducted based on the physical environment in which it operates, and its historical experience and culture. We have briefly examined how navies, air forces, and armies have approached operations during and immediately after the Cold War. We then discussed how the CF has approached operations during this same period.

The concept of NCW originated with certain Western navies, and therefore NCW's basic principles were forged on the anvil of the Anglo-American tradition of naval command by the hammer of operations at sea. Despite the importance of these origins, they have not been explored in any detail; the description provided in chapter 3 is offered as an initial investigation of them. This investigation found that the two key building blocks of NCW were in place by the early 1990s, namely the Composite Warfare Commander (CWC) concept, which flattened traditional notions of command at sea, and the increasing use of satellite communications, which allowed real-time global communications at reliable data flow-rates. The Canadian Navy's practical experience in both domains and its unique relationship with the US Navy put it in a position to influence both the theory and practice of the nascent NCW concept. The participation of the Canadian Navy over the past two decades in a number of war games and exercises at sea was one aspect of its influence, but arguably its most important contribution to NCW came during post-9/11 operations, especially Operation Enduring Freedom (known in Canada as Operation Apollo). The experience of four successive Canadian commodores as CWC for the Arabian Sea theatre of operations (ultimately as commander task force 151) gave the Canadians unprecedented command experience in networked coalition operations. This experience is now being analyzed and put to use in subsequent activities.

The Canadian naval experience has indicated that future operations at sea will likely be composed of ad hoc "coalitions of the willing." The differences in culture, technology, and capabilities in such coalitions seem to indicate that c^2 in coalition operations will be redefined as "cooperation and coordination." This new paradigm of cooperation and coordination depends more on leadership or influence behaviours among peers than on traditional concepts of command involving the exercise of authority over subordinates. Therefore, in coalition operations the leadership concepts of emergent leadership and distributed leadership may be more useful than concepts of authority. In fact, one might characterize the high reputation that senior Canadian naval officers have earned in certain operational command positions as a type of emergent leadership based on three subclasses of personal power (i.e., expert, referent, and connection), rather than on position power.

Another product of the Canadian naval experience is the conclusion that while NCW is revolutionary in its implications for military operations, it is best managed in an evolutionary fashion in its development. The Canadian Navy's practice of using "mission fits" for a series of ships deploying on operations in succession allowed it to increase its capability for networked operations by making incremental equipment upgrades and, just as important, by improving knowledge of how to use that equipment throughout the fleet. Using this "mission fit" methodology, the Canadian Navy has been able to upgrade its fleet in a steady and fiscally manageable way, albeit in a way that did not ensure standardization across the fleet, especially between the two coasts. A key lesson learned here, at least for medium-sized navies, is that because of the different defence cycles (e.g., technology, experience, professional military education, and acquisition) NEOps should not be managed as a capital project but implemented as an evolutionary process that takes the various cycles into account.

Navies, therefore, perceive systems like NCW primarily as a command and control architecture necessary to effectively coordinate complex, multi-threat maritime operations over a wide area (generally within a theatre of operations, but up to global in scope). Within the naval C² framework, individuals are connected via their consoles, but they operate as elements of larger systems, such as the various ships' operations rooms within the fleet framework. Unlike the army model, therefore, where command is delegated down to the lowest possible levels, in navies the notion of mission command (or command-by-influence) might extend down only to the captain of a vessel and a very few of his specially delegated principal officers. Nevertheless, human-centred networks are the basis of the Canadian naval command style, and this style has proven itself to be particularly effective in recent coalition operations where it has been necessary to engage a wide variety of coalition members in a task force and then ensure their effective participation in operations. Canadian naval commanders therefore appreciate that information technology is an enabler, not an end in itself.

If navies have built the foundation of NCW, air forces have proposed an alternate structure, or paradigm, for military operations. While now included under the NCW umbrella, EBO has a distinct and different history from NCW. EBO's roots go back to the 1920s, when air power theorists like Douhet asserted that the long stalemate at sea and on land of the First World War could be overcome in future wars by air forces with their largely unfettered ability to attack enemy centres of gravity. Douhet's modern-day disciples have expanded his ideas in more formal expositions of EBO, but the basic concept remains the same – planning and taking actions, ranging from threats to bombing attacks, to cause changes in opponents' behaviour.

While some in the US Air Force still present EBO largely as a targeting exercise, more sophisticated variants of EBO have now been incorporated into NCW, joint doctrine, and other concepts like

Rapid Decisive Operations and Shock and Awe. Like NCW itself, the notion of EBO is still poorly understood and subject to different interpretations; as we know, there is no universally accepted theory of EBO, and at least six different variants now coexist. Furthermore, EBO's critics note that in the past it has not lived up to its promises. They point out that the chaotic nature of warfare makes it almost impossible to predict, with the accuracy necessary to achieve the results EBO enthusiasts have claimed, how various actions will achieve the desired second- and third-order effects, especially those associated with human behaviour.

Even if EBO is not a fully functional theory of war, Western air forces have adopted it as the guiding principle for integrating air operations into joint operations. Building on their experiences in the Gulf War in the early 1990s, Western air forces have created an elaborate C^2 system based on the Air Operations Centre to coordinate all aspects of air operations for the joint force commander. However, the air force approach to C^2 is largely incompatible with some of the C^2 concepts now being articulated in NCW policy documents, particularly self-synchronization, self-organization, and mission command or command-by-influence.

Given the nature of complex air operations, while there may be limited opportunities for self-synchronization and command-by-influence processes, for the foreseeable future air forces will rely on command-by-plan to execute their missions. However, in some circumstances – for example, when decisions could have enormous political consequences, such as the release of nuclear weapons or the shooting down of civilian aircraft – air forces will likely continue to use command-by-direction. And because of the type of C^2 structures air forces require to coordinate all aspects of a complex air campaign, the idea of a self-organizing system seems implausible. Therefore, because of the environment in which they operate, air forces today and in the foreseeable future will likely rely on command-by-plan and sometimes command-by-direction rather than command-by-influence. For these reasons, among others, the US Marine Corps has rejected EBO as a philosophy to guide its operations, because the Marine Corps philosophy is based on command-by-influence and human-centric networks.

Despite the attempt of some advocates of NCW to portray EBO as part of the theory of NCW, proponents of EBO might argue that NCW should be seen as an enabler of EBO, not vice versa. The experience of air forces is that networks are required to enable their operations, but the network should not be the primary consideration. In fact, from an air force perspective, the focus of NCW on inputs (the network) as opposed to outcomes (as proposed by EBO theories) could be seen as a step backward on the path to progress in theories of war.

Like airmen, soldiers after the First World War searched for solutions to the deadlock of the trenches. The solution favoured by most Western armies was manoeuvre, which by the end of the twentieth century had become almost an obsession with many of them. In recent times the American Army version of operational art has been adapted by most Western armies as the preferred method to implement manoeuvre warfare.

Although NCW-based transformation gained prominence as the driving concept in US military transformation at the beginning of the twenty-first century, operational art was in many ways the dominant paradigm in US military thought in the last fifteen years of the twentieth century. It was the intellectual basis for the creation of the US worldwide regional command system, so that the world could be divided into "theatres of operation" and campaigns could be planned and conducted on the operational-level geographical principles favoured by armies.

In an attempt to rectify its mistakes in Vietnam, the US Army engineered the renaissance of a nineteenth- and early twentieth-century European approach to war based on operational art. The legacy of this renaissance is considerable, as operational art remains the foundation of most Western joint doctrine today. Embedded in this doctrine are the notions of the geographical division of military effort into theatres of war where separate campaigns can be planned and executed, a campaign design process that uses the operational level of war to link tactical actions to strategic goals, and a reliance on extensive written doctrine as the basis for operational art. This approach has been described as an "objectives- based approach to operations" as opposed to an "effects-based approach to operations." In addition, this US Army school of thought has also inextricably linked the notion of manoeuvre on the battlefield with success at the operational level. Finally, implicit in this approach to war is the belief that the decisive actions of war are conducted by land forces.

However, the army concept of manoeuvre on the battlefield is not always compatible with the notion of manoeuvre at sea or

manoeuvre as understood by air forces; therefore, the word "manoeuvre" still invokes different mental pictures in different war-
fighting communities. Recently Boyd's OODA loop model has been
adopted by some in the US Army and elsewhere as an example of
manoeuvre in the "fourth dimension" of time. Aspects of the OODA
loop model have also been used by advocates of NCW to argue for
increased speed in decision-making cycles or "speed of command."
As many critics have noted, however, faster decisions do not necessarily mean better decisions. As with other theories of war, there is
still much work to be done before the OODA loop model can be said
to provide a precise guide to transformation in networked operations.

To counter the dominant land-centric influence on joint doctrine
based on manoeuvre and the operational-art concept, air forces and
navies have challenged many of the US Army interpretations of
these concepts. They devised their own concepts to explain how
operations should be conducted, at the very least in their respective
physical environments, and put forward their challenges with ideas
like EBO and NCW. This intellectual ferment has resulted in at least
five different approaches to operational art.

Besides service differences in approaches to operational art, there
are also national differences It is only recently, however, that some
progress has been made in Canada towards examining operational-
art concepts from a theoretical and doctrinal point of view, because
there is no sound intellectual base in this country for studies of
operational art, and an arbitrary bureaucratic process was used to
import mostly American Army ideas on the operational art into
Canadian doctrine. This situation has created a fragmented
approach to operational thought in this country, which explains
why the CF does not always follow the tenets of prevailing Western
doctrine.

In Canada (as in the US), the Army has the most coherently articulated theoretical approach to operational art, although recent
practical experience has led to a somewhat unique development.
Peace support operations in an alliance or coalition context have
been some of the most formative influences on the Canadian practice of operational art. Therefore, as we have seen, unlike many
other militaries, the Canadian Army's perception of the operational
level of war is not focused on operational manoeuvre or operational logistics, nor is it tied to a theatre of war. Rather Canadian

commanders seek to coordinate operational-level systems appropriate to a multi-agency environment and to use the force structures under their command to achieve operational-level objectives.

The Canadian Army's approach to operational art is based on these premises: land operations are complex and continuous, regularly involving physical and psychological isolation and more often than not, until too late, unseen lethality; adversaries are often non-state and motivated by issues other than that of policy; these adversaries attack in unpredictable ways using the strengths of an opponent as a weakness to gain a temporary advantage that can be exploited; conflict is not confined to discernible regions and involves all aspects of the spectrum of conflict in a specific area, including the disappearance of the distinction between combatants and non-combatants; and effective joint, multinational, and multi-agency operations are the key to success in future operations. Therefore, the objective, or end state, of the Canadian manoeuvrist approach to operational art is to impede the enemy's ability to conduct warfare as a cohesive force. In this vision, manoeuvre warfare should be based on these principles: focus on enemy vulnerabilities, not ground; avoid enemy strength and attack his weaknesses; focus on the main effort; and be agile.

Given their circumstances, successful practitioners of Canadian manoeuvre philosophy have had to place much greater emphasis on the creation of shared awareness in order to overcome the ambiguity and corresponding friction found in current operational environments. One way of creating this shared awareness is by commanders at all levels using what the Canadian Army calls mission command or command-by-influence, sometimes called "trust leadership."

The Canadian Army has explored how technology can facilitate its version of manoeuvre warfare since the 1990s. Terms such as digitization, NCW, network-centric operations, Network-Enabled Capability, and NEOps have all been used to define this fluid relation between operations and emerging technology. However, whatever terminology has been used, Canada's Army sees itself as a doctrine-based organization that uses technology to increase its capability to practice manoeuvre warfare. Therefore, for the Army, ideas of NCW or NEOps do not constitute a fundamental change in the manner in which it will conduct operations, but reflect a need to use emerging technology in a way that will support the user and enable manoeuvre doctrine so that Canada's small professional army can operate in today's complex environment of conflict. Networks created by the Canadian Army show that it views human, not technical, factors as the primary consideration in their creation. Furthermore, because of financial and other limitations, these networks are hybrid in nature, where the technology used is not always the latest variant, but only what is required to carry out the mission. In this situation command is predicated on communication, dissemination of intent, creation of shared awareness, and decentralized decision making.

NEOps has not yet been formally accepted as a principle supporting the transformation of DND, nor has it been clearly defined, but recent NEOps conceptual statements indicate a similarity to NCW in that NEOps is expected "to generate increased combat power by networking sensors, decision makers and combatants to achieve shared battlespace awareness, increased speed of command, higher operational tempo, greater lethality, increased survivability, and greater adaptability through rapid feedback loops.'"[1] However, a number of Canadian commentators note that NEOps is more focused on human factors than is NCW. There is also an awareness among many Canadian commentators that any definition of NEOps should be consistent with Canadian culture and ethos. DND and the CF should, therefore, be careful about borrowing a concept that may not be compatible with their needs and be cognizant of the fact that implementing NEOps will require more than simply overlaying a networking capability onto an existing organizational or command and control structure. Perhaps most importantly, from a Canadian point of view and based on recent Canadian experience, using NEOps in the JIMP context will require network architects not only to consider the mere use of information technology as an enabler, but also to address the much more complex issue of the creation of effective social networks.

Chapter 6 provided a representative sampling of some critiques of NCW based on different paradigms of war and other operations. Most of these critiques are related to the fact that NCW is a technology-centred approach to war fighting and other operations.

Biddle notes that new technical systems, like those proposed by NCW, can require very costly investments, but he reminds us that training and readiness require sizable investments as well. And without adequate investments in training and readiness, new technology will not necessarily be the force multiplier or decisive element that some believe it will be. Biddle's arguments have important implications for Canada and other similar countries, since potential US coalition partners must consider how to balance their investments among numbers and quality of personnel and quantities of sophisticated equipment.

A more fundamental criticism of NCW was put forward by Kagan, who asserted that its origins in 1990s business and technical processes were not necessarily conducive to a twenty-first century theory of war. The idea that in using NCW a military can achieve information dominance over an enemy in much the same way that some successful corporations have used information to dominate their markets is a dubious proposition at best, according to some critics, since unlike customers, enemies will usually try to frustrate attempts to gather intelligence, especially using the technical means favoured by NCW.

Even if NCW is able to fuse information into a common operating picture, Kolenda argues that the education, culture, and personalities of those viewing the picture will result in diverse interpretations of what is presented. Furthermore, a number of commentators have noted that the more efforts that are taken to standardize both the information and the interpretation of that information, the more likely it is that creativity and originality will be stifled. This suppression of creativity and originality will work against the development of command-by-influence in an NCW environment. To ensure that commanders are able to make optimal use of networked C² systems, it has been suggested that they should be within what Pigeau and McCann postulate as the "Balanced Command Envelope," so that the required symmetry amongst the competency, authority, and responsibility necessary for effective command is achieved. Furthermore, it has been argued that information technology will not guarantee self-synchronization in an NCW environment if commanders at all levels do not have the attributes required to do their jobs.

Another critique of the technical focus of NCW is its use of state-of-the-art remote sensing to gather information. Recent reports from Iraq contradict official NCW policy documents about the usefulness of this information, as users in the field support those critics who contend that to effect political transformation, an intimate "stand-in" connection with the culture and people you propose to transform is required.

The physical and cultural settings in which armed forces operate are the base for other critiques of NCW. As noted in this volume, air forces operate in the least cluttered battlespace, and in these circumstances command-by-direction and command-by-plan are not only possible, but perhaps necessary. Armies, on the other hand, usually operate in the most complex and chaotic operating environment, and therefore Western armies have for the most part adopted the doctrine of mission command or command-by-influence so that decisions can, in theory, be taken by those closest to the situation, often down to the level of the individual soldier. Navies operate in an environment of medium complexity, compared with air forces and armies, and therefore most Western navies in the Anglo-American command tradition have identified the need for a c^2 system to effectively coordinate maritime operations in a relatively complex, multi-threat environment over a wide area (normally a theatre of operations, but one that could encompass global operations). Within the naval framework, although individuals would be connected via their consoles, they would be operating as elements of larger systems, such as the various ships' operations rooms (at the lowest level) within the fleet framework. These new c^2 systems, which are based on the original NCW concept, are enabling navies to replace the command-by-direction style practised by navies for centuries with a uniquely naval command-by-influence style that has increasingly been observed in naval operations of the Canadian Navy and some other navies in the Anglo-American command tradition in the last two decades.

Some recent Canadian concerns with networked operations, as articulated in the NCW documentation that appears to be informing the emerging NEOps concept in Canada, are as follows: increased speed of command could lead to a faster means to make old mistakes; increased mission effectiveness, usually defined in NCW documents in terms of defeating an enemy more efficiently, should be described in the context of a peacekeeping operation with phrases like "improving quality of life"; the implicit assumption in NCW and NEOps is that simply providing people with access to the same information will enable common understanding, but how "common intent" can actually be promoted among network players, who are often from diverse backgrounds and cultures (both national and organizational), represents a major challenge for the future; a major challenge in implementing NEOps will be its effect on organizational culture and structure, but there appears to be a lack of scientific investigation regarding NEOps and its impact on and interaction with CF culture; there appears to be a focus on the technical aspects of NEOps to the detriment of how transformation based on NEOps might affect the roles that Canada plays in international affairs and the impact that NEOps will have on the human dimension of CF operations.

A number of critics have noted that NCW's origins were in concepts designed to prevail in "big" wars but that today's environment is one of "small" wars. The most recent variant of "small war" theory, Fourth Generation Warfare (4GW), or idea-driven warfare, offers a number of challenges to the NCW concept. As we have seen, many of the technical aspects of NCW (especially its c^2 architecture) were designed to operate in a naval environment of medium complexity to deal with relatively well-known and quantifiable threats. The 4GW environment is highly complex, and many of the threats are unpredictable, difficult to quantify, and changing, since opponents use "4GW judo" to keep large Western security, military, and legal bureaucracies off balance. If one accepts Lind's assertion that the real centre of gravity of 4GW opponents is a shared religious or ideological goal, where common purpose and zealotry replace recognizable military command structures and military pattern equipment, then NCW technology-based systems may not have the flexibility to deal with this threat.

The Canadian military experience in the post-Cold War environment, particularly the Army's stabilization efforts in post-conflict Afghanistan and the Navy's command of coalition operations in the Arabian Sea, reinforces the belief that the human network, not the technical network, should be the basis for future approaches to CF transformation. The CF experience with the Canadian deployment with ISAF Rotation IV in 2003 provided lessons similar to those gleaned from Canada's mission to Bosnia during Operation Palladium as well as certain recent domestic operations. Typically, in this new environment the CF are sent into complex security environments without clear strategic objectives save for the imperative to establish a "secure environment."

In operating in these complex security environments, the primary difference between the NCW concept and the Canadian Army's practice in peace support operations since 1992 is that instead of having the technical network as the centre of focus, the Canadian Army has focused on creating human-centric networks, with technology of various kinds being adapted to meet the needs of the human network. This adaptation has often resulted in technical networks that are ad hoc and hybrid in nature, but this has been necessary because the Canadian Army has not had access to the full range of information management tools employed by US armed forces.

The Canadian Army's post-Cold War experience has demonstrated that the challenges posed by peace support operations in post-conflict situations are best met with holistic solutions that identify issues that must be addressed simultaneously, in a distributed fashion, across elements of national power. This methodology is best achieved with the human-centric, not the network-centric approach.

The lack of clear strategic objectives and a unifying concept of operations has in many cases forced the Canadian Army to exert great effort to construct these hybrid networks to allow for interaction and to create cooperation among the numerous diverse groups involved in a particular mission. The creation of a cohesive and focused plan that fosters unity of effort in these circumstances demands an understanding of the nature of societal reconstruction and renewal required for a particular region, but it also must be linked to structures that provide focus to all entities within a given country, like Afghanistan. The Canadian experience has been that it is often the military that must provide the assistance and impetus necessary for the formulation of an overarching system, or networked community, to achieve the ultimate objective of removing the causes of war in a region. The plan "Creating a National Economy: The Path to Security and Stability in Afghanistan" was described as an example of the product of the Canadian Army's human-centric network in Afghanistan – a coherent plan to achieve a strategic vision of a rejuvenated and independent Afghanistan as an integrated member of the global community.

The Canadian Navy's experience was somewhat different, given its very good access to developing US Navy network-enabled systems and procedures. Interestingly, however, that naval experience points to the same conclusion as the army experience, reinforcing the validity of an approach that balances the human and technological factors. This further demonstrates that a "one size fits all" concept may not be best suited to the unique capabilities required by each service, even in a joint and combined, and increasingly integrated, operating environment.

In summary, NEOps as a concept has a promising future if it is predicated on Canadian needs and culture. However, there is significant risk in placing too much reliance on concepts like NCW that put the technological cart before the human requirements that should drive any transformation initiative. Therefore, future development of the NEOps concept should be firmly rooted in the Canadian context and based on Canadian experience. NEOps concept development should be complemented by the relevant experience of others, but it should avoid slavishly copying other frameworks, as DND has sometimes done in the past. In the Canadian context of human-centred networks, research to support the development of the NEOps concept should be conducted in the areas related to the human dimension of networks based on theory and on Canadian practical experience. In this way, NEOps could become a suitable model to support the transformation of the CF and DND.

Notes

CHAPTER ONE

1 Bercuson, "Defence Education for 2000 ... and Beyond," 30.
2 Cebrowski and Garstka, "Network-Centric Warfare," 28–35; later expanded in Alberts, Garstka, and Stein, *Network Centric Warfare*.
3 Hughes, "'New Orthodoxy' Under Fire," 57.
4 Gregory, "From Stovepipes to Grids," 18.
5 US Department of Defense, Office of Force Transformation, *The Implementation of Network-Centric Warfare*, 3.
6 Babcock, "Canadian Network Enabled Operations Initiatives," 4.
7 US Department of the Navy, "FORCEnet," 1.
8 Ibid., 12.
9 The six dimensions are physical, information technology, data, cognitive, organizational, and operating. Ibid., 19–20.
10 Hughes, "'New Orthodoxy' under Fire," 57.
11 Finch, "Approaching Transformational Coalition Operations," 18.
12 English, "The Operational Art," 20.
13 Thomson and Adams, "Network Enabled Operations," 4–6.
14 The latest proposed official "definition" of NEOps is a rather long statement of intent and not a definition as such. See Canada, DND, "DND/CF Networked Enabled Operations: Keystone Document (Final Draft)," 10.
15 Babcock, "Canadian Network Enabled Operations Initiatives," 4.
16 Schneider, "Transforming Advanced Military Education," 15–22.
17 See, for example, Paret, ed., *Makers of Modern Strategy*, and Handel, *Masters of War*.
18 Cohen, "Neither Fools nor Cowards," A12.
19 Holley, "A Modest Proposal," 14–20.

20 Toffler and Toffler, *War and Anti-War*, as a follow-on to Toffler, *The Third Wave*.

21 Wattie, "Absent-minded Professor," 9.

22 Loren B. Thompson, chief operating officer of the Lexington Institute public policy think-tank and a national security specialist, cited in Hughes, "'New Orthodoxy' Under Fire," 57.

23 These are the definitions of "theory" selected by the authors of the latest official NCW policy statement: "A theory is 'a hypothesis assumed for the sake of argument or investigation, an unproved assumption.' It is also 'a formulation of apparent relationships or underlying principles of certain observed phenomena which has been verified to some degree.'" US Department of Defense, Office of Force Transformation, *The Implementation of Network-Centric Warfare*, 15.

24 Many translations of Clausewitz's *On War* exist, but the best-known modern edition is that edited and translated by Michael Howard and Peter Paret.

25 Young, "Clausewitz and His Influence," 14–15. Like Clausewitz, the Chinese writer on military strategy Sun Tzu is frequently quoted out of context to support or "validate" modern doctrine.

26 This definition of theory has been modified from *Funk & Wagnalls Canadian College Dictionary*, Toronto: Fitzhenry & Whiteside, 1986, 1389.

27 Shy, "Jomini," 143–85.

28 According to conventional wisdom, science is a strictly logical process. However, while objectivity is the essence of the scientist's attitude to his or her work, in the acquisition of knowledge scientists are guided not by logic and objectivity alone but also by such non-rational factors as rhetoric, propaganda, and personal prejudice. Therefore science should be considered not the guardian of rationality in society but merely one major form of its cultural expression. See, for example, Broad and Wade, *Betrayers of the Truth*.

29 Metz, "A Wake for Clausewitz," 126–32.

30 See Young, "Clausewitz and His Influence," 9–21, for a summary of this issue in a Canadian context.

31 Johnston, "Doctrine Is Not Enough," 30–9.

32 Corn, "World War IV," np.

33 Military theorists such as Shimon Naveh of the Department of History Tel Aviv University note that Kuhnian theory regarding the progress of science and the processes by which paradigms are replaced by the scientific community has application in the realm of warfare theory and his-

tory. See, for example, Naveh, *In Pursuit of Military Excellence*, xiii-xv. Kuhn argued that when, in the development of a natural science, an individual or group is able to produce a synthesis or a new form of the older theory, and is able to attract most of the next generation's practitioners, the older schools disappear. Kuhn, *The Structure of Scientific Revolutions*, 18–19.

34 Gat, *A History of Military Thought*, 824.

35 Murray, *The Making of Strategy*, 1, 7.

CHAPTER TWO

1 See, for example, Garnett, "The Canadian Forces," 5–10.

2 Szafranski, "Peer Competitors," 116.

3 Cohen, "A Revolution in Warfare," 52–3.

4 Baumann, "Historical Perspectives," np.

5 Kagan, "War and Aftermath," 3. Cebrowski died on 12 November 2005. Arkin, "Spiraling Ahead," 143.

6 Barnett, "The Seven Deadly Sins," 36–9. Barnett subsequently worked closely with Cebrowski in the US Department of Defense Office of Force Transformation. His book *The Pentagon's New Map: War and Peace in the Twenty-First Century* offers an interesting fusion of technology and human networks.

7 Liddell, "Operational Art," 50–5.

8 Andrew Krepinevich Jr, a leading commentator on military affairs and a member of the 1997 National Defense Panel, cited in Ricks and White, "Scope of Change," A06.

9 Cited in Ricks and White, "Scope of Change," A06.

10 Sloan, *The Revolution in Military Affairs*, 46–8.

11 Ricks and White, "Scope of Change," A06.

12 These ideas are discussed in detail in Barzelay and Campbell, *Preparing for the Future*.

13 DND, *Leadership in the Canadian Forces*, 10.

14 Pigeau and McCann, "Re-conceptualizing Command and Control," 53.

15 The concepts of the strategic corporal and mission command are discussed in a Canadian context in DND, *Duty with Honour*, 64; and DND, *Leadership in the Canadian Forces*, 124–5.

16 Czerwinski, "Command and Control at the Crossroads," 121–32.

17 See, for example, McCann, Pigeau, and English, "Analysing Command Challenges."

18 This section is excerpted from Sharpe and English, *Principles for Change*, 71–2.

19 Pigeau and McCann, "Re-conceptualizing Command and Control," 56.

20 Ross Pigeau and Carol McCann, "Re-conceptualizing Command and Control," presentation given to Command and Staff Course 31, Canadian Forces College, 3 September 2004.

21 This material is excerpted from Sharpe and English, *Principles for Change*, 78–80. The notion of common intent is developed in more detail in Ross Pigeau and Carol McCann, "Establishing Common Intent: The Key to Co-ordinated Military Action," in English, ed., *The Operational Art*, 85–108.

CHAPTER THREE

1 Soeters and Recht, "Culture and Discipline in Military Academies," 179, 183. This idea is discussed in more detail in English, *Understanding Military Culture*, chapters 4, 5, and 6.

2 For example: Oliver Warner, *Command at Sea: Great Fighting Admirals from Hawke to Nimitz* (New York: St Martin's, 1976); Jack Sweetman, ed., *The Great Admirals: Command at Sea, 1587–1945* (Annapolis, Md.: US Naval Institute Press, 1997). A very new addition to the literature promises to be more of the same: Michael A. Palmer, *Command at Sea* (Cambridge: Harvard University Press, 2005); reviewed by Vice Admiral James Stavridis (US Navy) in US *Naval Institute Proceedings* 131, no. 6 (June 2005), 83–4.

3 English et al., *Command Styles in the Canadian Navy*, 148–9.

4 Keegan, *The Mask of Command*, 10.

5 English et al., *Command Styles in the Canadian Navy*, 48–9.

6 DND, *Leadmark*.

7 For example, the standard survey work, Paret, ed., *Makers of Modern Strategy*, dedicates only one chapter out of twenty-eight to the strategy of sea power (on Mahan), with no mention elsewhere of such figures as Julian Corbett or Raoul Victor Castex.

8 Sumida, *Inventing Grand Strategy*, 25.

9 Crowl, "Alfred Thayer Mahan," 444–77.

10 P. Masson, "Alfred Thayer Mahan," entry in Corvisier, ed., *A Dictionary of Military History*, 478–9.

11 Hattendorf, *The Evolution of the U.S. Navy's Maritime Strategy*, 4–5.

12 Ibid., 20–2.

13 Hayward's opening statement to Congress appeared as "The Future of U.S. Sea Power," US *Naval Institute Proceedings*, May 1979, 66–71. Here it is quoted from Palmer, *Origins of the Maritime Strategy*, xv.

14 A.W.H. Pearsall, "Sir Julian Stafford Corbett," in Corvisier, ed., *A Dictionary of Military History*, 177–8.

15 See, for example, Goldrick and Hattendorf, *Mahan Is Not Enough*.

16 Clark, "Sea Power 21," 32–41.

17 Based upon a presentation by Colonel (retd) Brian MacDonald, "Are We There Yet?: The Interactive Cycle Problem," to a conference on The New Defence Agenda: Transforming National Defence Administration, sponsored by Defence Management Studies, Queen's University; the Institute for Research on Public Policy; the Canadian Defence Industries Association; and the Canadian Defence Associations Institute, Ottawa, 6 April 2005.

18 Kaplan, "How We Would Fight China," 49–64.

19 Lautenschläger, "Technology and the Evolution of Naval Warfare," 214. One of the authors of this book (Gimblett) recalls participating as a member of a squadron of Canadian destroyers attached to a US Navy task force that comprised three carrier battle groups sailing from Pearl Harbor in the winter of 1983. They were dispersed roughly 300 miles apart to swing by Kamchatka in an early demonstration of the Maritime Strategy.

20 A valuable analysis of command at sea in the battleship era is Gordon, *Rules of the Game*. Although no similar account exists for the postwar era of carrier aviation, Friedman, *The Postwar Naval Revolution* provides a useful context.

21 English et al., *Command Styles in the Canadian Navy*, 85.

22 Carr, "Network Centric Coalitions," 5.

23 The description that follows is necessarily simplified in the interests of making the general principles accessible to lay readers. Naval professionals will know that the development of the concept was neither as smooth nor as effectively put into practise as is suggested here.

24 The CWC concept was originally promulgated in the secret allied compilation of experimental tactics as EXTAC 740.

25 US Navy, *Multinational Maritime Operations Doctrine Manual*, 3–21, 3–22.

26 Carr, "Network Centric Coalitions"; and Geraghty, "Will Network-Centric Warfare Be the Death Knell." Mitchell, "Small Navies and Network-centric Warfare," was a seminal article on the subject, but it was published just as Canadian command of task force 151 in the Arabian Sea was proving the concept already was being achieved in practice.

27 Previous tactical datalinks (also known as TADILs) had passed teletype messages that needed to be hand-plotted. Link-11 transmitted computer graphics to be automatically overlaid on the ship-generated radar display. The standard present-day TADIL is Link-16, which incorporates embedded "pages" of amplifying information on each graphic symbol.

28 Vice-Admiral (retired) Lynn Mason, interview with Dr Richard Gimblett, Halifax, NS, 12 May 2005.

29 Grove, *Battle for the Fiords* is an account of the NATO Exercise Teamwork 1988, including a good description of the part played by the Canadian Task Group (CATG).

30 Miller and Hobson, *The Persian Excursion.*

31 Gimblett, "MIF or MNF?," 193.

32 Background on the development of JOTS through the follow-on system of JMCIS (Joint Maritime Command Information System) to the current GCCS-M can be found at www.fas.org/irp/program/core/jmcis.htm (Federation of American Scientists, Intelligence Resource Program, Intelligence Programs and Systems). Accessed 7 February 2007.

33 It is impressive just how little computing power is required to run the ship's basic systems: the networks described in the text are still in service nearly two decades after initial design, with those for the Halifax class about to be updated as part of their mid-life upgrade (nothing similar is envisaged for the Iroquois class, since those 40-year-old vessels are to be withdrawn over the coming decade).

34 Commander Dan Stovel, interview with Dr Richard Gimblett, Halifax, NS, 13 May 2005.

35 Orincon Corporation, www.orincon.com/techfinder/project_detail. cfm?key_project=64&key_subcategory=35.

36 Interview, Mason.

37 They were *Regina* (1997), *Ottawa* (1998), *Regina* (1999), *Calgary* (2000), *Charlottetown* (2001, first half) and *Winnipeg* (2001, latter half). All were Esquimalt-based ships, except for *Charlottetown*; another Halifax-based ship, *Toronto*, was detached from a deployment with the Standing Naval Force Atlantic in mid-1998 to lend additional support to coalition forces in assuring access of UNSCOM to Iraqi sites. For enthusiastic American assessments of the frigate integrations, see Crockett, "Professional Notes," 65–7; and Stavridis, "They Got Game," 51–4.

38 Lieutenant-Commander (retired) R.W.H. (Bill) McKillop, e-mail correspondence with Dr Richard Gimblett, 1 June 2005.

39 Morse and Thomas, "STANAVFORLANT," 61–4.

40 Cebrowski and Garstka, "Network-Centric Warfare: Its Origin and Future."

41 Hay and Gile, *Global War Game.*

42 Captain (N) (retired) Ian Parker, interview with Dr Richard Gimblett, Ottawa, 31 May 2005.

43 For an overview of Operation Apollo, with particular emphasis on the naval participation, see Gimblett, *Operation Apollo.*

44 Fox, *Iraq Campaign 2003*, provides an overview of the Royal Navy's role.

45 Goldrick, "In Command in the Gulf," 38–41.

46 Commodore (retired) Eric Lerhe, e-mail correspondence with Dr Richard Gimblett, 27 June 2005.

47 Commodore Roger Girouard, interview with Dr Richard Gimblett, at sea (on board HMCS *Iroquois*, Arabian Sea), 28 May 2003.

48 Commodore (retired) Eric Lerhe, interview with Dr Richard Gimblett, Halifax, NS, 13 May 2005.

49 DND, *Leadmark*, 163–4.

50 Maddison, "The Canadian Navy's Drive for Trust and Technology."

51 LCdr (US Navy) Steven Loeffler, interview with Dr Richard Gimblett, at sea (on board HMCS *Algonquin*, off Vancouver Island), 25 May 2005.

52 English et al., *Command Styles in the Canadian Navy*, 111, 151.

53 Johnson, "Net-centric Fogs Accountability," 32.

54 Ibid., 35.

55 Ibid., 32.

56 Ibid., 33.

57 The general subject of accountability as it applies to the Canadian Forces has been a very important issue, particularly as a result of several incidents in the 1990s. The concept of accountability requires that responsibilities be clearly stated and appropriately promulgated to ensure that they will be fulfilled. Also, these actions ensure that in the event an individual fails to fulfil his or her responsibilities, he or she can be held to account. In September 1999 the second edition of "Organization and Accountability – Guidance for Members of the Canadian Forces and Employees of the Department of National Defence" was published under the authority of the chief of the Defence Staff and the deputy minister of the department. It is available at www.forces.gc.ca/site/minister/eng/authority/OA_e.htm (National Defence and the Canadian Forces). Accessed 4 March 2007.

58 Johnson, "Net-centric Fogs Accountability," 34–5.

59 English et al., *Command Styles in the Canadian Navy*, 151.

60 The conclusions suggested in this paragraph are an amalgam of observations derived from interviews with a variety of sailors of all ranks (from master seaman through chief petty officer 2nd class, and lieutenant through commodore) serving in and recently retired from the Canadian Navy, over the course of research at the Canadian Forces Maritime Warfare Centre, Halifax, 12–13 May 2005, and at sea on board HMCS *Algonquin*, 24–27 May 2005. Without assigning formal attribution, the authors are indebted to the observations provided.

61 DND, "Employment of Collaboration at Sea."

62 "Information Management Director – Terms of reference," e-mail to Dr Richard Gimblett from Lt(N) Hal Shiels, Canadian Forces Fleet School Halifax, 21 June 2005.

63 Free, "Network-Centric Leadership," 58–60.

CHAPTER FOUR

1 Leonard, "Learning from History," 267–303.

2 US Army, *Operations*, 2–19.

3 Vallance, *The Air Weapon*, 91, cited in Hallion, "Airpower and the Changing Nature of War," 42.

4 US Air Force, *Air Force Basic Doctrine*, 13.

5 Owens, "Reshaping Tilted against the Army?"

6 US Air Force, *Air Force Basic Doctrine*, 2.

7 Westrop, "Aerospace Doctrine Study."

8 Warden, *The Air Campaign*.

9 Greer, "Operational Art for the Objective Force," 26. Greer is a former director of the US Army's School of Advanced Military Studies (SAMS).

10 Harris, "Effects-Based Operations," iii.

11 A detailed account of this example can be found in Mandeles et al., *Managing "Command and Control,"* especially 1–8.

12 Heide, "Canadian Air Operations," 79.

13 One of the most detailed descriptions of EBO has been written by former US Air Force officer Maris McCrabb, "Effects-based Coalition Operations," in Tate, ed., *Proceedings of the Second International Conference*, 134–46. A summary of the air force view of the evolution of Effects-Based Operations can be found in Meilinger, "The Origins of Effects-Based Operations," 116–22.

14 See, for example, Ho, "The Advent of a New Way of War," 3–4.

15 See Segre, "Giulio Douhet," 351–66, for a summary of Douhet's theories.

16 These issues are discussed in detail in Phillip S. Meilinger, ed. *The Paths of Heaven: The Evolution of Airpower Theory*. Maxwell, Ala.: Air University Press, 1997, chapters 1–8.

17 Ho, "The Advent of a New Way of War," 5–10.

18 There is an extensive literature on this topic. See, for example, Segre, "Giulio Douhet"; Pape, *Bombing to Win*; and Park, "'Precision' and 'Area' Bombing."

19 See, for example, Schmitt, "Command and (out of) Control," 55–8; Hall, "Decision Making," 28–32; and Rousseau, "Command in a Complex Battlespace," 55–83.

20 Crowder, "Effects-Based Operations," 16–25.

21 Ullman and Wade, *Shock and Awe*.

22 Excerpt from the introduction to Ullman and Wade, *Shock and Awe*, quoted in "Shock and Awe: The Idea Behind the Buzzwords," *Washington Post*, B03.

23 Kagan, "War and Aftermath," 5–8.

24 Kolenda, "Transforming How We Fight," 100–21.

25 Ho, "The Advent of a New Way of War," 23–4.

26 Mann et al., "Dominant Effects: Effects-Based Joint Operations," 92–100.

27 US Marine Corps, "Effects Based Operations Conference," Powerpoint presentation dated 7 September 2005, in the possession of the authors.

28 The fixed-wing bombers were Gothas and *Riesenfluzeug* (Giants). The Giants had 6 engines and a wingspan of 140 feet, carried a crew of 9, had a maximum bomb load of 2 tonnes, and could fly 600 miles non-stop. Fredette, *The Sky on Fire*, 6–7.

29 Crabtree, *On Air Defense*, 32–3; and English, *Marching through Chaos*, n8, 141.

30 Crabtree, *On Air Defense*, 31–2.

31 Ibid., 212.

32 The best description of the US Air Force's theoretical evolution from a cultural perspective is Builder, *The Icarus Syndrome*. See also Winton, "A Black Hole in the Wild Blue Yonder," 32–42, for the lack of a theoretical focus in the US Air Force before the Gulf War.

33 Fought, "The Tale of the C/JFACC," 10–11.

34 US Department of Defense, Office of Force Transformation, *The Implementation of Network-Centric Warfare*, ii.

35 Pratt, "A Clash of Service Doctrines."

36 Roddy, "Network-Centric Operational Warfare," 9.

37 For example, decisions related to a recent intrusion of a civilian aircraft into restricted air space over Washington, D.C., involved the US Secretary of Defense. Hsu and Mintz, "Military Was Set to Down Cessna," AOI.

38 Davis, "Centralized Control/Decentralized Execution," 98–9.

39 Ibid., 98.

40 Crowder, "Effects-Based Operations," 16–25.

41 US Navy, "FORCEnet," 12.

42 Fought, "The Tale of the C/JFACC," 10–11.

CHAPTER FIVE

1 US Army, *Operations*, 2–5.

2 The term "regional C-in-C" has been replaced by "Unified Commander," but certain commanders are still given responsibility for their assigned parts of the globe. For more details and a map of how the responsibility for various parts of the world have been assigned to certain unified commanders, see www.defenselink.mil/specials/unifiedcommand/ (US Department of Defense, "Unified Command Plan"). Accessed 4 March 2007.

3 Naveh, *In Pursuit of Military Excellence*, 251.

4 Starry, "A Perspective on American Military Thought," 6–10.

5 Attrition warfare can be best expressed as "a toe-to-to slugging match in which each side assumes the other will abide by predictable rules and that sheer weight of numbers and material will determine the outcome." Fallows, *National Defense*, 26–7.

6 McAndrew, "Operational Art and the Northwest European Theatre of War 1944," 20.

7 Menning, "Operational Art's Origins," 42.

8 Peskett, "Levels of War," 101–2.

9 See Geyer, "German Strategy," 527–97; and Bond and Alexander, "Liddell Hart and DeGaulle," 598–623 for the origins of post-First World War manoeuvre doctrine.

10 Naveh, *In Pursuit of Military Excellence*, 251, 329.

11 "A Statement on the Posture of the United States Army – Fiscal Year 1996" (Washington, D.C.: Dept of the Army, 1995), 26–7, cited in Hallion, "Airpower and the Changing Nature of War," 42. As late as the spring of 2000 this was still the US Army's view; see, for example, Aubin, "Stumbling towards Transformation," 39–47.

12 US Army, *Operations*, 2–6.

13 "Heretical" is used here in the sense of beliefs or opinions contrary to established doctrine or opinion.

14 Grant, "Closing the Doctrine Gap," 52.

15 Robert H. Thomas and Richard Gimblett summarize more of the naval approach to these issues in their bibliography "Maritime Doctrine at the Operational Level of War," which is available on the Canadian Forces College Information Resource Centre website at http://wps.cfc.forces. gc.ca/bibliographies/173_en_maritime.htm. Accessed 4 March 2007.

16 Hughes, "Naval Maneuver Warfare," p. 13 of online edition.

17 Tritten, "Naval Doctrine ... From the Sea," 1–2.

18 Grant, "Closing the Doctrine Gap," 48–52.

19 See the US Joint Electronic Library, www.dtic.mil/doctrine/service_publications_navy_pubs.htm. Accessed 16 June 2005.

20 Tritten, "Naval Doctrine ... From the Sea," 7, 14.

21 Grant, "Closing the Doctrine Gap," 50; and Sloan, "The United States and the Revolution in Military Affairs," 9.

22 DND, *Leadmark*, 31.

23 Leonhard, "Factors of Conflict," pp. 1–2 of online version.

24 Greer, "Operational Art for the Objective Force," 26.

25 Morton, *A Military History of Canada*, 200. The detailed evolution of Murray's command in the context of its development of Canadian joint doctrine has recently been described in Goette, "The Struggle for a Joint Command and Control System." See also Milner, "Rear-Admiral Leonard Warren Murray," 7–8.

26 William McAndrew cited in Peskett, "Levels of War," 102.

27 Coombs, "Perspectives on Operational Thought," 81, 85, 89–90.

28 Gimblett, *Operation Apollo*, 108–12, 114, 120–2, 134–6.

29 These ideas are a product of discussions held by an informal Operational Art Working Group sponsored by Canadian Forces College, the Canadian Forces Leadership Institute, and Defence Research and Development Canada – Toronto. The details of the work of this group can be found in Coombs, "Leadership, Command and Operational Art."

30 DND, *A Role of Pride and Influence in the World*.

31 English, *Marching through Chaos*, 167; and Polk, "A Critique of the Boyd Theory," 257–8.

32 Polk, "A Critique of the Boyd Theory," 257–60. For more on Boyd's ideas, see also Hammond, "From Air Power to Err Power."

33 Polk, "A Critique of the Boyd Theory," 271.

34 US Department of Defense, Office of Force Transformation, *The Implementation of Network-Centric Warfare*, 5, 7.

35 Forgues, "Command in a Network-Centric War," 29.

36 Polk, "A Critique of the Boyd Theory," 258.

37 Ibid., 265, 267, 270. See English, *Understanding Military Culture*, especially chapters 4 and 6 for more detail on dysfunctional aspects of US Army culture.

38 Polk, "A Critique of the Boyd Theory," 272; and Fadok, "John Boyd and John Warden," 388–9.

39 For example, the highest casualty rates suffered by the Canadian Corps occurred during the mobile warfare phase of the last "Hundred Days" of the war. The 45,830 casualties it incurred between Amiens and Mons represented 20 per cent of the Canadian Expeditionary Force total for the entire war, and was more than the 44,735 casualties suffered by the Canadian Army in the Northwest Europe campaign in the Second World War. In addition, in the Second World War, the mobile campaigns on the Eastern Front claimed more lives than all the other theatres of that war combined. English, *Marching through Chaos*, 62–4.

40 Peace support operations refer to the gamut of peace operations, from traditional peacekeeping, which requires lightly armed forces to observe and report violations of a previously brokered agreement, to peace enforcement, which requires robust forces capable of imposing and maintaining a cessation of conflict.

41 Michael Wyly believes the two key principles of manoeuvre warfare to be understanding of the higher commander's intent and understanding of the designation of the main effort, or "*Schwerpunkt.*" Wyly, "Teaching Maneuver Warfare," 257–8.

42 DND, *Canada's Army: We Stand on Guard for Thee*, 86–7. It has been argued that the history of the Canadian land force in the twentieth century does not lend itself to the institutionalization of the philosophy required to affect a manoeuvre-type doctrine and related concepts such as that delineated in *We Stand on Guard for Thee*. The experiences of the Boer War, First and Second World Wars, Korea, the Cold War, and peacekeeping, as well as the interludes between major conflicts, have produced a vision of conflict that seems to rely on a symmetrical vision of the battlefield. This perspective is predicated on an understanding of conflict as an orderly and progressive series of engagements, battles, and campaigns that will result in victory. However, it can also be opined that Canadian land operations from the early 1990s onwards have prompted a re-evaluation of past lessons to embrace the reality of the present – an asymmetric, chaotic environment where decisions are made in minimal time under less than ideal circumstances. In this setting the precepts of manoeuvre warfare, like the principle of subsidiarity, are reinforced.

43 Jarymowycz, "Doctrine and Canada's Army," 50.

44 Strange, *Centers of Gravity and Critical Vulnerabilities*, ix.
45 Oliviero, "Response to 'Doctrine and Canada's Army,'" 140.
46 We are grateful to Colonel (retired) Eric MacArthur for providing us with this summary. A more complete version can be found in DND, *Canada's Army*, 100–5.
47 DND, *Conduct of Land Operations*.
48 DND, *Land Force Tactical Doctrine*, vol. 2, pp. 1–4 to 1–5.
49 Polk, "A Critique of the Boyd Theory," 266.
50 Johnston, "The Myth of Manoeuvre Warfare," 27–8.
51 Hope, "Manoeuvre Warfare and Directive Control," 10.
52 Ibid., 10.
53 Simpkin, *Race to the Swift*, 230.
54 Richard Simpkin describes these qualities as the parameters of command. Ibid., 229. Dr James Schneider has further delineated the psychological domain of command as the ability of commanders to impose their will in order to carry an idea through the planning phase to successful execution while overcoming the inherent friction of combat. Schneider views the physical domain of war as the process of destruction of an army's ability to operate within the cybernetic and moral realms of conflict, and thus disrupting the cohesion of the organization. Schneider, *The Theory of Operational Art*, 6–7.
55 Through kinetic or high-intensity types of operations, one attempts to overwhelm the enemy through all means available (but primarily violence), and the focus is normally on the destruction of the enemy.
56 This was also the case in Operation Desert Storm, as noted in Hope, "Changing a Military Culture," 5.
57 Hillen, "Peace(keeping) in Our Time," 17–34.
58 Echevarria and Biever, "Warfighting's Moral Domain," 3–6.
59 Lind et al., "The Changing Face of War," 3–11.
60 Myatt, "Comments on Maneuver," 40–2.
61 Echevarria and Biever, "Warfighting's Moral Domain."
62 DND, *The Debrief the Leaders Project (Officers)*, 26.
63 DND, *1994 White Paper on Defence*, 21.
64 Ibid., 49–50.
65 Office of the Commander, Land Force Command, "Army 2000 Campaign Plan," (file no. 3136-1 (Comd), dated 21 March 1996), 2.
66 DND, *Conduct of Land Operations*, 9.
67 DND, *Land Force Information Operations*, 7–12.
68 Analogous to the United States Army Training and Doctrine Command (TRADOC), Land Force Doctrine and Training System provides, albeit in a

rudimentary form, the same impetus for innovation in the Canadian Army that occurred in the US Army.

69 DND, *Future Army Development Plan*, 13, 37.

70 DND, *The Future Security Environment*, 52 –4.

71 DND, *Intelligence, Surveillance, Target Acquisition, and Reconnaissance*, 1 and C-1.

72 Garnett, "The Canadian Forces," 5–10; and Forgues, "Command in a Network-Centric War," 23–30.

73 "Technology, led by digitization, will present the opportunity for a quantum leap in command support capabilities. Rather than simply automating the reporting system, technology must be used to expedite the time and energy consuming tasks of collection, analysis and presentation. The focus must be on developing intelligent and rugged decision support systems. The potential exists to develop human to machine interfaces that will allow the commander and staff to literally decide and act with the speed of thought." DND, *Command, Sense, Act, Shield and Sustain*, 17.

74 DND, *The Future Security Environment*, 40–1.

75 DND, *Command, Sense, Act, Shield and Sustain*, 18.

76 DND, *Operations in the Expanded Battlespace*, 3 and 10. Mission Command "places the emphasis on decentralizing authority and empowering personal initiative." DND, *Land Force Command*, i.

77 DND, *Evaluation of the Impact of LCF215*; and DND, *Defence Research and Development Canada*.

78 DND, *The Force Employment Concept for the Army*, 4.

79 Terminology variations were noted in e-mail correspondence from DLSC 6 to Howard G. Coombs, 13 May 2005. For further discussion of the Canadian Army and current concepts of NCW, see the proceedings of the 2003 Army Symposium, DND, *Towards the Brave New World*; and DND, *Future Force Concepts*.

80 A view counter to this argument and supporting Cebrowski and Garstka is contained in Sloan, *The Revolution in Military Affairs*, 123–42.

81 McAndrew, "Soldiers and Technology," 20–4.

82 McAndrew, "Operational Art and the Canadian Army's Way of War," 90.

83 Van Creveld states, "From Plato to NATO, the history of command in war consists essentially of an endless quest for certainty – certainty about the state and intention's of the enemy forces; certainty about the factors that together constitute the environment in which war is fought, from the weather and the terrain to radioactivity and the presence of chemical warfare agents; and, least but definitely not least, certainty about the

state, intentions and activities of one's own forces." Van Creveld, *Command in War*, 264.

84 Van Creveld, *Command in War*, 265.

85 Ibid., 269–70.

86 E-mail from an American officer in Iraq to Howard G. Coombs, 28 May 2005.

87 Simpkin, *Race to the Swift*, 241.

88 Hope, "Manouvre Warfare and Directive Control," 8–9.

89 Van Creveld, *Command in War*, 269–70.

90 DND, "National Defence Strategic Capability Investment Plan."

91 Babcock, "Canadian Network Enabled Operations Initiatives," 3.

92 Thomson and Adams, "Network Enabled Operations," 5.

93 Ibid., 7–8.

94 Babcock, "Canadian Network Enabled Operations Initiatives," 4.

95 DND, "DND/CF Networked Enabled Operations."

CHAPTER SIX

1 Sloan, *The Revolution in Military Affairs*, 149.

2 For example, Patrick Pidgeon, e-mail on personal experience in Afghanistan, 27 November 2006. In the authors' possession.

3 Biddle, "Victory Misunderstood," 140, 162, 178–9.

4 Owens, "Reshaping Tilted against the Army?"

5 Leonhard, "Factors of Conflict," p. 2 of online edition.

6 "Network Centric USW – Exploring the Realities."

7 Barnett, "The Seven Deadly Sins," 38–9.

8 Lescher, "Network-centric: Is It Worth the Risk?," 60, 62–3.

9 Kagan, "War and Aftermath," 6.

10 Kolenda, "Transforming How We Fight," 103, citation from 114.

11 Hughes, "'New Orthodoxy' under Fire," 57.

12 Ignatius, "Standoffish Soldiering," A15.

13 Quote from an e-mail from the operations officer of a task force in Iraq, 26 May 2005.

14 Hughes, "'New Orthodoxy' under Fire," 57.

15 Van Creveld, *Command in War*, 9, 262–3.

16 Polk, "A Critique of the Boyd Theory," 272.

17 DND, "Vectors 2020," 14–15.

18 Van Creveld, *Command in War*, 9, 262–3.

19 Forgues, "Command in a Network-Centric War," 23–30.

20 See Johnson, "Net-centric Fogs Accountability."

21 Forgues, "Command in a Network-Centric War," 23–30.
22 Cebrowski, in US Department of Defense, Office of Force Transformation, *The Implementation of Network-Centric Warfare*, 1.
23 Babcock, "Canadian Network Enabled Operations Initiatives," 4.
24 Thomson and Adams, "Network Enabled Operations," 10.
25 Ibid., 12.
26 Ibid., 13–14.
27 Ibid., 15. The authors are referring to H.R. McMaster, "Crack in the Foundation: Defense Transformation and the Underlying Assumption of Dominant Knowledge in Future War," unpublished US Army War College paper, 2003, and to L. Warne et al., "The Network Centric Warrior: The Human Dimension of Network Centric Warfare" (DSTO Report CR–0373) (Edinburgh, South Australia: Defence Systems Analysis Division, Defence Science and Technology Organisation, Australian Department of Defence, 2004). MacNulty is cited on 50–1.
28 Ibid., 15–16.
29 Ibid., 16. The authors are referring to C. McCann and R.A. Pigeau, eds. *The Human in Command: Exploring the Modern Military Experience* (New York: Kluwer Academic/Plenum Publishers, 2000).
30 Ibid., 16.
31 Ibid., 17. The authors are referring to Canada, DND, *Canadian Forces Strategic Operating Concept* (Ottawa: National Defence Headquarters, 2004).
32 Ibid., 19.
33 Sloan, *The Revolution in Military Affairs*, 152–3.

CHAPTER SEVEN

1 See, for example, McKenzie, "Elegant Irrelevance," 51–60.
2 Lind et al., "The Changing Face of War," 2–11.
3 Lind, "All War All the Time."
4 Ibid.
5 Wilson et al., "4GW – Tactics of the Weak Confound the Strong."
6 Van Creveld, "High Technology," 61–4.
7 *The Concise Oxford Dictionary*, 6th ed., 477. For descriptions of variations on "small wars," see Porch, "Bugeaud, Galliéni, Lyautey," 408–43; and Shy and Collier, "Revolutionary War," 815–62. For another view and commentary on more recent "small wars," see Corum and Johnson, *Airpower in Small Wars*.
8 Wilson et al., "4GW – Tactics of the Weak Confound the Strong."

9 Observations regarding the creation of networked operations in Afghanistan are derived from Coombs and Hiller, "Planning for Success."

10 Fuller believed the ultimate weakness of Clausewitzian theory to be its misunderstanding of the role that peace played in shaping warfare and that halting the violence of conflict, when disconnected from the strategy required for the establishment of a lasting peace, results in nothing more than a temporary cessation of hostilities. Fuller, *The Conduct of War*, 76.

11 Noonan, "Foreign Policy Research Institute E-Notes."

12 The expression "three-block war" has been attributed to General Charles Krulak, a former commandant of the United States Marine Corps, and indicates that in the current operational environment militaries will be confronted with a span of tactical challenges ranging from humanitarian to peace support activities to combat within a short period of time and in the space of three city blocks. Krulak, "The Strategic Corporal."

13 The authors of this paper are indebted to a number of Canadian Forces officers, including Commander (N) Chris Henderson, Lieutenant Colonels Allen Black (now retired) and Ian Hope, as well as Majors Cathy Amponin and Mike Pepper, both of the United States Air Force, and the staff of ISAF V whose work in Afghanistan is contained throughout this chapter.

14 E-mail to Howard G. Coombs from an American officer, Afghanistan, July 2003.

15 Pearson, "Nobel Lecture."

16 For further study concerning concepts of nation rebuilding, see Pugh, *Regeneration of War-Torn Societies*.

17 The most recent publication of Canadian peace support operations doctrine lists conflict prevention, peacemaking, peace building, traditional peace keeping operations, complex peace keeping operations, enforcement actions, and humanitarian operations as the various categories of peace support activities. DND, Peace Support Operations, 2–3 to 2–5. An historical perspective of the challenges experienced by Canadians during peace support operations is contained in Morton, *A Military History of Canada*, 4th ed., 277–81.

18 It should be noted that while in retrospect one can see UNEF I as a "traditional" or customary form of peacekeeping, for the participants it was a distinctly new type of military mission with protocols that had to be developed as the operation matured. See Swettenham, *Some Impressions of UNEF*.

19 Morton, *A Military History of Canada*, 4th ed., 242.

20 Granatstein, *Canada and Peace-keeping Operations*.

21 Pratt, "The Way Ahead for Canadian Foreign and Defence Policy."

22 This incident involved elements of the First Battalion The Royal Canadian Regiment Battle Group. A detailed discussion of these violent events, ostensibly precipitated by United Nations High Commission on Refugees (UNHCR) resettlement of Serbians to what had become a primarily Croatian community, and their aftermath, is contained in Swain, *Neither War nor Not War*, 1-25. Defence analyst David Rudd has reinforced the necessity of the partnerships of the 3D concept with: "The security provided by robust, well-equipped military forces in strife-torn lands opens the door to political reconstruction, which begets economic and social development, which in turn reinforces security." Rudd, "Canada's New Defence Policy."

23 Pearson, "Nobel Lecture."

24 Cox, "Major General Andrew Leslie," 8.

25 "The ATA replaced the Afghan Interim Authority (AIA). In accordance with the Bonn Agreement, the ATA organised a Constitutional Loya Jirga in late 2003 to pave the way for the election of an Afghan government by early 2004." Because of a number of factors the election was delayed until October 2004 and resulted in the inauguration of President Karzai that December. See Strand et al., "Peacebuilding in a Regional Perspective."

26 Diagram from Hope, "A Strategic Concept."

27 The Transitional Islamic State of Afghanistan, Ministry of Finance Consultation Draft, "National Priority Programs (NPPs): An Overview," 23/24 June 2004, 3.

28 NATO, ISAF, "Creating a National Economy: The Path to Security and Stability in Afghanistan," June 2004.

29 "Security Issues Dampen Afghan Investments."

30 Afghanistan Ministry of Finance Consultation Draft, "Securing Afghanistan's Future: Accomplishments and the Strategic Path Forward," 29 January 2004.

31 From Hope, "A Strategic Concept."

32 The term "regional power-broker" is used to denote an individual who controls power that is neither sanctioned nor necessarily authorized by the state.

33 United Nations Organization on Drugs and Crime and the Government of Afghanistan Counter Narcotics Directorate, *Afghanistan Opium Survey 2003*.

34 From Hope, "A Strategic Concept."

35 A senior Uzbek commander expressed the following sentiment: "If there is any law to be implemented, it should first be done in Kabul, then in the provinces, and then in the districts. If you don't disarm the defence ministry, which is dominated by Panjshiris, people will ask, 'Why are you disarming this province or that province?' If the top four commanders [in Kabul] place their arms in the hands of central government, people like us will have no problems placing our arms in the centre." Quoted in International Crisis Group, "Disarmament and Reintegration in Afghanistan."

36 Canada, *The Force Employment Concept for the Army*, 4–5.

37 US Department of Defense, Office of Force Transformation, *Implementation of Network-Centric Warfare*, 4–5 (emphasis in original).

38 Guttman, "The Shadow Pentagon."

39 CFC-A had the mission of conducting "full spectrum operations throughout the area of operations in order to establish enduring security, defeat Al Qaida/Taliban and deter the re-emergence of terrorism in Afghanistan." Untitled ISAF Liaison Officer Document (1 September 2004) in possession of Howard G. Coombs.

40 In the quest for certainty, commanders sometimes use qualified and trusted officers to act as observers and report their findings. These special agents exist outside the chain of command and report back to the originating authority in the manner of a telescope directed towards a certain point. These officers provide information from specified units and operations. Griffin, *The Directed Telescope*, 1.

41 Address by Zinni, "How Do We Overhaul?"

CHAPTER EIGHT

1 Thomson and Adams, "Network Enabled Operations," 5.

Bibliography

Alberts, David S., John J. Garstka, and Frederick P. Stein. *Network Centric Warfare: Developing and Leveraging Information Superiority.* US Department of Defense, C4ISR Cooperative Research Program, 1998.

Arkin, William A. "Spiraling Ahead: With the Loss of its Greatest Champion, What's To Become of Transformation?" *Armed Forces Journal International* 143 (February 2006), 39–42.

Aubin, Stephen. "Stumbling towards Transformation." *Strategic Review* 28, no. 2 (Spring 2000), 39–47.

Babcock, Sandy. "Canadian Network Enabled Operations Initiatives." Ottawa: National Defence Headquarters, Directorate of Defence Analysis, nd (2004?).

Barnett, Thomas P. *The Pentagon's New Map: War and Peace in the Twenty-First Century.* New York: G.P. Putnam's Sons, 2004.

– "The Seven Deadly Sins of Network-Centric Warfare." *US Naval Institute Proceedings* 125, no. 1 (January 1999), 36–9.

Barzelay, Michael and Colin Campbell. *Preparing for the Future: Strategic Planning in the US Air Force.* Washington, D.C.: Brookings Institution Press, 2003.

Baumann, Robert F. "Historical Perspectives on Future War." *Military Review* 76, no. 2 (March–April 1997), 40–8.

Bercuson, D.J. "Defence Education for 2000 ... and Beyond." In *Educating Canada's Military*, report of a workshop held at the Royal Military College of Canada, 7–8 December 1998.

Biddle, Stephen. "Victory Misunderstood: What the Gulf War Tells Us about the Future of Conflict." *International Security* 21, no. 2 (Fall 1996), 139–79.

Bond, Brian and Martin Alexander. "Liddell Hart and DeGaulle." In Paret, ed., *Makers of Modern Strategy*, 598–623.

Broad, William and Nicholas Wade. *Betrayers of the Truth*. New York: Simon and Schuster, 1982.

Builder, Carl H. *The Icarus Syndrome: The Role of Air Power Theory in the Evolution and Fate of the US Air Force*. London: Transaction Publishers, 1994.

Canada, Department of National Defence. *1994 White Paper on Defence*. Ottawa: Canada Communications Group, 1994.

– *Canada's Army: We Stand on Guard for Thee*. B-GL-300-000/FP-00 (1 April 1998).

– *Command, Sense, Act, Shield and Sustain: DLSC Report 01/01 Future Army Capabilities*. Kingston: Fort Frontenac, Directorate of Land Strategic Concepts, January 2001.

– *Conduct of Land Operations – Operational Level Doctrine for the Canadian Army*. B-GL-300-001/FP-000 (1 July 1998).

– *The Debrief the Leaders Project (Officers)*. Ottawa: Office of the Special Advisor to the Chief of the Defence Staff for Professional Development, May 2001. Available at www.cda.forces.gc.ca/2020/engraph/research/debrief/doc/DebriefLeaders_e.PDF. Accessed 4 March 2007.

– *Defence Research and Development Canada – Concept of Operations for Collaborative Working (Version 2.2)* (25 May 2004).

– "DND/CF Networked Enabled Operations: Keystone Document (Final Draft) – A DND/CF Concept and Roadmap Paper." Ottawa: National Defence Headquarters, 30 May 2005.

– *Duty with Honour: The Profession of Arms in Canada*. Kingston: Canadian Defence Academy, 2003.

– "Employment of Collaboration at Sea Systems in Task Group Operations." CFCD 106D, TACNOTE 6520 (Unclassified), March 2003.

– *Evaluation of the Impact of LFC2IS on Warfighting Operations at the Brigade Group Level: Operational Research – Army Experiment 6B – EXERCISE ATHENE WARRIOR*. Kingston: Fort Frontenac, Directorate of Land Strategic Concepts, November 2002.

– *The Force Employment Concept for the Army*. Kingston: Fort Frontenac, Directorate of Land Strategic Concepts, 2004.

– *Future Army Development Plan*. Kingston: Fort Frontenac, Directorate of Land Strategic Concepts, 8 March 1999.

– *Future Force Concepts for Future Army Capabilities*. Kingston: Fort Frontenac, Directorate of Land Strategic Concepts, 2003.

- *The Future Security Environment:* DLSC *Report 99–2.* Kingston: Fort Frontenac, Directorate of Land Strategic Concepts, August 1999.
- *Intelligence, Surveillance, Target Acquisition, and Reconnaissance (ISTAR):* DLSC *Army Experimentation Centre Report 9906 Army Experiment 1.* Kingston: Fort Frontenac, Directorate of Land Strategic Concepts, December 1999.
- *Land Force Command.* B-GL-300-003/FP-000, 21 July 1996.
- *Land Force Information Operations.* B-GL-300-005/FP-001, 18 January 1999.
- *Land Force Tactical Doctrine, Vol. 2.* B-GL-300-002/FP-000, 16 May 1997.
- *Leadership in the Canadian Forces: Conceptual Foundations.* Kingston: Canadian Defence Academy, 2005.
- *Leadmark: The Navy's Strategy for 2020.* Ottawa: Chief of Maritime Staff, 2001. Available at: www.navy.dnd.ca/leadmark/doc/index_e.asp. Accessed 4 March 2007.
- "National Defence Strategic Capability Investment Plan, Issue 1." November 2003). Available at www.vcds.forces.gc.ca/dgsp/pubs/rep-pub/ddm/scip/intro_e.asp. Accessed 4 March 2007.
- *Operations in the Expanded Battlespace:* DLSC *Future Army Experiment.* Kingston: Fort Frontenac, Directorate of Land Strategic Concepts, June 2001.
- *Peace Support Operations.* B-GJ-005-307-FP-030, 6 November 2002.
- *A Role of Pride and Influence in the World: Defence Policy Statement.* Ottawa: National Defence Headquarters, Minister of National Defence, April 2005. Available at www.forces.gc.ca/site/Reports/dps/main/toc_e.asp. Accessed 4 March 2007.
- *Towards the Brave New World: Canada's Army in the 21st Century.* Kingston: Fort Frontenac, Directorate of Land Strategic Concepts, 2003.
- "Vectors 2020." Unpublished National Defence Headquarters staff paper, Director Air Strategic Plans, dated 5 April 2002.
- Carr, James. "Network Centric Coalitions: Plug, Pass, or Plug-in?" Unpublished US Naval War College paper, 1999.
- Cebrowski, Arthur K. and John J. Garstka. "Network-Centric Warfare: Its Origin and Future." US *Naval Institute Proceedings* 124, no. 1 (January 1998), 28–35.
- Clark, Vern. "Sea Power 21: Projecting Decisive Joint Capabilities." US *Naval Institute Proceedings* 128, no. 10 (October 2002), 32–41.
- Clausewitz, Carl. *On War.* Michael Howard and Peter Paret, eds. and trans. Princeton: Princeton University Press, 1976.

Cohen, Eliot A. "Neither Fools nor Cowards." *Wall Street Journal*, 13 May 2005, A12.
– "A Revolution in Warfare." *Foreign Affairs* 75, no. 2 (March/April 1996), 37–54.
Coombs, Howard G. "Leadership, Command and Operational Art Project Progress Report – March 2003 to March 2004." Report written for the CF Leadership Institute, available at the Information Resource Centre, Canadian Forces College.
– "Perspectives on Operational Thought." In English et al., eds., *The Operational Art*, 75–96.
Coombs, Howard G. and Rick Hiller. "Planning for Success: The Challenge of Applying Operational Art in Post-Conflict Afghanistan." *Canadian Military Journal* 6, no. 3 (Autumn 2005), 5–14.
Corn, Tony. "World War IV as Fourth-Generation Warfare." *Policy Review* 135 (February and March 2006), np. Available at www.hoover.org/publications/policyreview/4868381.html. Accessed 4 March 2007.
Corum, James S. and Wray R. Johnson. *Airpower in Small Wars: Fighting Insurgents and Terrorists*. Lawrence, Kan: University Press of Kansas, 2003.
Corvisier, André, ed. *A Dictionary of Military History and the Art of War*. English edition edited by John Childs. London: Blackwell, 1994.
Cox, Jim. "Major General Andrew Leslie – Kabul & ISAF." *Frontline* Issue 4 (September/October 2004). Available at www.frontline-canada.com/Defence/archives/2004Issues.html#ISSUE4. Accessed 4 March 2007.
Crabtree, James D. *On Air Defense*. Westport, Conn.: Praeger, 1994.
Crockett, Michael. "Professional Notes: O Canada!" *US Naval Institute Proceedings* 124, no. 12 (December 1998), 65–7.
Crowder, Gary L. "Effects-Based Operations: The Impact of Precision Strike Weapons on Air Warfare Doctrines." *Military Technology* 27, no. 6 (June 2003), 16–25.
Crowl, Philip A. "Alfred Thayer Mahan: The Naval Historian." In Paret, ed., *Makers of Modern Strategy*, 444–77.
Czerwinski, Thomas J. "Command and Control at the Crossroads." *Parameters* 26, no. 3 (Autumn 1996), 121–32.
Davis, Mark G. "Centralized Control / Decentralized Execution in the Era of Forward Reach." *Joint Force Quarterly*, no. 35 (Autumn 2004), 95–9.

Echevarria, Antulio J. and Jacob D. Biever. "Warfighting's Moral Domain." *Military Review* 90, no. 2 (March–April 2000), 3–6.

English, Allan. *Understanding Military Culture: A Canadian Perspective.* Montreal and Kingston: McGill-Queen's University Press, 2004.

English, Allan D., ed. *Air Campaigns in the New World Order.* University of Manitoba, Winnipeg: Centre for Defence and Security Studies, 2005.

– *The Changing Face of War.* Montreal and Kingston: McGill-Queen's University Press, 1998.

– *The Operational Art: Canadian Perspectives – Leadership and Command.* Kingston: Canadian Defence Academy Press, 2006.

English, Allan, et al. *Command Styles in the Canadian Navy.* Defence Research and Development – Toronto, Contract Report CR 2005–096, 31 January 2005.

English, Allan, et al., eds. *The Operational Art: Canadian Perspectives – Context and Concepts.* Kingston: Canadian Defence Academy Press, 2005.

English, John. "The Operational Art: Developments in the Theories of War." In McKercher and Hennessy, eds., *The Operational Art*, 7–27.

English, John A. *Marching through Chaos: The Descent of Armies in Theory and Practice.* Westport, Conn.: Praeger, 1996.

Fadok, David S., "John Boyd and John Warden." In Phillip S. Meilinger, ed., *The Paths of Heaven: The Evolution of Airpower Theory.* Maxwell, Ala.: Air University Press, 1997, 357–98.

Fallows, James. *National Defense.* New York: Random House, 1981.

Finch, D.P. "Approaching Transformational Coalition Operations along the Standardization, Interoperability and Integration (SI2) Continuum." Draft CCRTS Paper 024, J7 Doctrine and Standardization, Director General Joint Force Development, Department of National Defence, nd (2005).

Forgues, Pierre. "Command in a Network-Centric War." *Canadian Military Journal* 2, no. 2 (Summer 2001), 23–30.

Fought, Stephen. "The Tale of the C/JFACC: A Long and Winding Road." *RAF Air Power Review* 7, no. 4 (Winter 2004), 10–11.

Fox, Robert. *Iraq Campaign 2003: Royal Navy and Royal Marines.* London: Agenda Publishing, 2003.

Fredette, Raymond. *The Sky on Fire: The First Battle of Britain and Birth of the Royal Air Force.* Washington, D.C.: Smithsonion Institution Press, 1966.

Free, Jennifer. "Network-Centric Leadership: Why Trust is Essential." *US Naval Institute Proceedings* 131, no. 6 (June 2005), 58–60.

Friedman, Norman. *The Postwar Naval Revolution: Warships, Weapons and Policies – The Navies Which Might Have Evolved, and the Factors Which Shaped Their Actual Development.* Annapolis, Md.: Naval Institute Press, 1986.

Fuller, J.F.C. *The Conduct of War, 1789–1961: A Study of the Impact of the French, Industrial, and Russian Revolutions on War and Its Conduct.* New Brunswick, NJ: Rutgers University Press, 1961; reprint Cambridge and New York: Da Capo Press, 1992.

Garnett, Gary. "The Canadian Forces and the Revolution in Military Affairs: A Time for Change." *Canadian Military Journal* 2, no. 1 (Spring 2001), 5–10.

Gat, Azar. *A History of Military Thought.* New York: Oxford University Press, 2001.

Geraghty, Barbara A. "Will Network-Centric Warfare Be the Death Knell for Allied/Coalition Operations?" Unpublished US Naval War College paper, 1999.

Geyer, Michael. "German Strategy in the Age of Machine Warfare, 1914–45." In Paret, ed., *Makers of Modern Strategy*, 527–97.

Gimblett, Richard. "MIF or MNF? The Dilemma of the 'Lesser' Navies in the Gulf War Coalition." In Michael L. Hadley et al., eds., *A Nation's Navy: In Quest of Canadian Naval Identity.* Montreal and Kingston: McGill-Queen's University Press, 1996.

– *Operation Apollo: The Golden Age of the Canadian Navy in the War Against Terrorism.* Ottawa: Magic Light, 2004.

Goette, Richard. "The Struggle for a Joint Command and Control System in the Northwest Atlantic Theatre of Operations." Unpublished MA thesis, Queen's University, 2002.

Goldrick, James. "In Command in the Gulf." *US Naval Institute Proceedings* 128, no. 12 (December 2002), 38–41.

Goldrick, James and John B. Hattendorf. *Mahan Is Not Enough: The Proceedings of a Conference on the Works of Sir Julian Corbett and Sir Herbert Richmond.* Newport, RI: Naval War College, 1993.

Gordon, Andrew. *Rules of the Game: Jutland and British Naval Command.* London: John Murray, 1996.

Granatstein, J.L. *Canada and Peace-keeping Operations: Report No. 4, Directorate of History.* Ottawa: Department of National Defence, 22 October 1965.

Grant, Rebecca. "Closing the Doctrine Gap." *Air Force Magazine* 80, no. 1 (January 1997), 48–52.

Greer, James K. "Operational Art for the Objective Force." *Military Review* 82, no. 5 (September–October 2002), 22–9.

Gregory, Bill. "From Stovepipes to Grids." *Armed Forces Journal International* 136, no. 6 (January 1999), 18–19.

Griffin, Gary B. *The Directed Telescope: A Traditional Element of Effective Command*. Fort Leavenworth, Kan.: Combat Studies Institute, 1991.

Grove, Eric with Graham Thompson. *Battle for the Fiords: NATO's Forward Maritime Strategy in Action*. Annapolis, Md.: Naval Institute Press, 1991.

Guttman, Dan. "The Shadow Pentagon." The Center for Public Integrity. Washington, DC, 29 September 2004. Available at www.publicintegrity.org/pns/report.aspx?aid=386. Accessed 4 March 2007.

Hall, John D. "Decision Making in the Information Age: Moving beyond the MDMP Military Decision-making Process." *Field Artillery* (September–October 2000), 28–32.

Hallion, Richard. "Airpower and the Changing Nature of War." *Joint Forces Quarterly* (Autumn/Winter 1997–98), 39–46.

Hammond, Grant T. "From Air Power to Err Power: John Boyd and the Opponent's Situational Awareness." In Peter W. Gray and Sebastian Cox, eds., *Air Power Leadership: Theory and Practice*. London: The Stationery Office, 2002, 107–28.

Handel, Michael I. *Masters of War: Classical Strategic Thought*, 3rd ed. London: Frank Cass, 2001.

Harris, John T. "Effects-Based Operations: Tactical Utility." Unpublished MA thesis, US Army Command and General Staff College, 2004.

Hattendorf, John B. *The Evolution of the US Navy's Maritime Strategy, 1977–1986*, Newport Paper No. 19. Newport, RI: US Naval War College, 2004.

Hay, Bud and Bob Gile. *Global War Game: The First Five Years*, Newport Paper No. 4. Newport, RI: Naval War College, 1993.

Heide, Rachel Lea. "Canadian Air Operations in the New World Order." In Allan D. English, ed., *Air Campaigns in the New World Order*, 77–92.

Hillen, John. "Peace(keeping) in Our Time: The UN as a Professional Military Manager." *Parameters* 26, no. 3 (Autumn 1996), 17–34.

Ho, Joshua. "The Advent of a New Way of War: Theory and Practice of Effects Based Operations." Singapore: Institute of Defence and Strategic Studies, Working Paper no. 57, December 2003.

Holley, I.B. "A Modest Proposal: Making Doctrine More Memorable." *Airpower Journal* 9, no. 4 (Winter 1995), 14–20.

Hope, Ian. "Changing a Military Culture: Manouvre Warfare and a Canadian Operational Doctrine, Part 1 of 2." *The Canadian Land Force Command and Staff College Quarterly Review* 5, no. 1/2 (Spring 1995), 1–6.

– "Manoeuvre Warfare and Directive Control: The Basis for a New Canadian Military Doctrine, Part 2 of 2." *The Canadian Land Force Command and Staff College Quarterly Review* 5, no. 1/2 (Spring 1995), 7–15.

– "A Strategic Concept for the Development of Afghanistan." Unpublished International Security Assistance Force presentation, June 2004.

Hsu, Spencer S. and John Mintz. "Military Was Set to Down Cessna: Authority Granted as Plane Strayed Deep into Capital." *Washington Post*, 25 May 2005, A01.

Hughes, David. "'New Orthodoxy' under Fire." *Aviation Week & Space Technology* (29 September 2003), 57–8.

Hughes, Wayne P. "Naval Maneuver Warfare." *Naval War College Review* 50, no. 3 (Summer 1997), 25–49. Available at www.nwc.navy.mil/press/Review/1997/summer/art2su97.htm. Accessed 4 March 2007.

Ignatius, David. "Standoffish Soldiering." *Washington Post*, 5 August 2003, A15.

International Crisis Group. "Disarmament and Reintegration in Afghanistan." International Crisis Group Report No. 65. Kabul/Brussels, 30 September 2003.

Jarymowycz, Roman J. "Doctrine and Canada's Army." *The Army Doctrine & Training Bulletin* 2, no. 3 (August 1999).

Johnson, Chris. "Net-centric Fogs Accountability." US *Naval Institute Proceedings* 129, no. 5 (May 2003), 32–5.

Johnston, Paul. "Doctrine Is Not Enough: The Effect of Doctrine on the Behavior of Armies." *Parameters* 30, no. 3 (Autumn 2000), 30–9.

– "The Myth of Manoeuvre Warfare: Attrition in Military History." In English, ed., *The Changing Face of War*, 22–32.

Kagan, Frederick W. "War and Aftermath." *Policy Review Online* 102 (August/September 2003). Available at www.hoover.org/publications/policyreview/3448101.html. Accessed 4 March 2007.

Kaplan, Robert. "How We Would Fight China: The Next Cold War." *The Atlantic Monthly*, June 2005, 49–64.

Keegan, John. *The Mask of Command*. London: Jonathan Cape, 1987.

Kolenda, Christopher D. "Transforming How We Fight: A Conceptual Approach." *Naval War College Review* 56, no. 2 (Spring 2003), 100–21.

Krulak, Charles C. "The Strategic Corporal: Leadership in the Three Block War." *Marine Corps Gazette* 83, no. 1 (January 1999), 18–22.

Kuhn, Thomas S. *The Structure of Scientific Revolutions*, 2nd ed. Chicago: University of Chicago Press, 1970.

Lautenschläger, Karl. "Technology and the Evolution of Naval Warfare." In Steven E. Miller and Stephen Van Evera, eds., *Naval Strategy and National Security: An International Security Reader*. Princeton: Princeton University Press, 1988.

Leonard, Raymond W. "Learning from History: Linebacker II and US Air Force Doctrine." *Journal of Military History* 58 (April 1994), 267–303.

Leonhard, Robert R. "Factors of Conflict in the Early 21st Century." *Army Magazine* 50, no. 1 (January 2003), 31–5. Available at : www.ausa.org/webpub/DeptArmyMagazine.nsf/byid/CCRN–6CCS72. Accessed 4 March 2007.

Lescher, William K. "Network-centric: Is It Worth the Risk?" *US Naval Institute Proceedings* 125, no. 7 (July 1999), 58–63.

Liddell, Daniel E. "Operational Art and the Influence of Will." *Marine Corps Gazette* 82, no. 2 (February 1998), 50–5.

Lind, William S. "All War All the Time: The Military Game Has Changed and the US Isn't Ready." *San Franciso Chronicle*, 6 July 2003, D1.

Lind, William S., et al. "The Changing Face of War: Into the Fourth Generation." *Military Review* 69 (October 1989), 2–11.

McAndrew, William. "Operational Art and the Canadian Army's Way of War." In McKercher and Hennessy, eds., *The Operational Art*, 87–102.

– "Operational Art and the Northwest European Theatre of War 1944." *Canadian Defence Quarterly* 21, no. 3 (December 1991), 19–26.

– "Soldiers and Technology." *The Army Doctrine and Training Bulletin* 2, no. 2 (May 1999), 20–4.

McCann, C., R. Pigeau, and A. English. "Analysing Command Challenges Using the Command and Control Framework: Pilot Study Results." Technical Report, Defence Research and Development

Canada – Toronto no. TR-2003-034, 1 February 2003. Available at
http://pubs.drdc-rddc.gc.ca/pubdocs/pcow1_e.html. Accessed 4 March
2007.

McCrabb, Maris. "Effects-based Coalition Operations: Belief, Framing
and Mechanism." In Austin Tate, ed., *Proceedings of the Second
International Conference on Knowledge Systems for Coalition
Operations*, 23–24 April 2002, Toulouse, France, 134–46. Available at
www.aiai.ed.ac.uk/project/coalition/ksco/ksco-2002/pdf-parts/
S-ksco-2002-paper-02-mccrabb.pdf. Accessed 4 March 2007.

McKenzie, Kenneth F., Jr. "Elegant Irrelevance: Fourth Generation
Warfare." *Parameters* 23, no. 3 (Autumn 1993), 51–60.

McKercher, B.J.C. and Michael A. Hennessy, eds. *The Operational Art:
Developments in the Theories of War*. Westport, Conn.: Praeger,
1996.

Maddison, Paul. "The Canadian Navy's Drive for Trust and Technology
in Network-Centric Coalitions: Riding Comfortably Alongside, or
Losing Ground in a Stern Chase?" Unpublished paper prepared for
Advanced Military Studies Course 7, Canadian Forces College, 2004.
Available at http://wps.cfc.dnd.ca/papers/amsc/amsc7/maddison.htm.
Accessed 4 March 2007.

Mandeles, Mark D., et al. *Managing "Command and Control" in the
Persian Gulf War*. Westport, Conn.: Praeger, 1996.

Mann, Edward, et al. "Dominant Effects: Effects-Based Joint
Operations." *Aerospace Power Journal* 15, no. 3 (Fall 2001), 92–100.

Meilinger, Phillip S. "The Origins of Effects-Based Operations." *Joint
Force Quarterly* 35 (October 2004), 116–22.

Menning, Bruce W. "Operational Art's Origins." *Military Review* 77, no.
5 (September–October 1997), 32–47.

Metz, Steven. "A Wake for Clausewitz: Toward a Philosophy of
21st-Century Warfare." *Parameters* 24, no. 4 (Winter 1994–95),
126–32.

Miller, Duncan (Dusty) E. and Sharon Hobson. *The Persian Excursion:
The Canadian Navy in the Gulf War*. Clementsport, NS: Canadian
Institute of Strategic Studies, 1995.

Milner, Marc. "Rear-Admiral Leonard Warren Murray: Canada's Most
Important Operational Commander." In Michael Whitby et al., eds.,
*The Admirals: Canada's Senior Naval Leadership in the Twentieth
Century*. Toronto: Dundurn Press, 2006, 97–124.

Mitchell, Paul T. "Small Navies and Network-centric Warfare: Is There a
Role?" *Naval War College Review* 56, no. 2 (Spring 2003), 83–99.

Morse, David and Douglas Thomas. "STANAVFORLANT under Canadian Command." *Canadian Military Journal* 1, no. 2 (Summer 2002), 61–4.

Morton, Desmond. *A Military History of Canada*, 4th ed. Toronto: McClelland and Stewart Ltd., 1999.

Murray, Williamson. *The Making of Strategy: Rulers, States and War.* Cambridge: Cambridge University Press, 1994.

Myatt, J. Michael. "Comments on Maneuver." *Marine Corps Gazette* 82, no. 10 (October 1998), 40–2.

Naveh, Shimon. *In Pursuit of Military Excellence: The Evolution of Operational Theory*. London and Portland: Frank Cass, 1997.

"Network Centric USW – Exploring the Realities." *Semaphore: Newsletter of the Sea Power Centre, Australia*, Issue 12 (November 2004).

Noonan, Michael P. "Foreign Policy Research Institute E-Notes: The Military Lessons of Operation Iraqi Freedom." Available at www.fpri.org/enotes/20030501.military.noonan.militarylessonsiraqifreedom.html. Accessed 4 March 2007.

Oliviero, Chuck. "Response to 'Doctrine and Canada's Army – Seduction by Foreign Dogma: Coming to Terms with Who We Are.'" *The Army Doctrine and Training Bulletin* 2, no. 4 (Winter 1999), 140–1.

Owens, Mackubin Thomas. "Reshaping Tilted against the Army?" *The Washington Times*, 24 November 2002.

Palmer, Michael A. *Command at Sea*. Cambridge: Harvard University Press, 2005.

– *Origins of the Maritime Strategy: American Naval Strategy in the First Postwar Decade*. Washington, D.C.: Naval Historical Center, 1988.

Pape, Robert A. *Bombing To Win: Air Power and Coercion in War.* Ithaca, NY: Cornell University Press, 1996.

Paret, Peter, ed., *Makers of Modern Strategy*. Princeton: Princeton University Press, 1986.

Park, W. Hays. "'Precision' and 'Area' Bombing: Who Did Which, and When?" *Journal of Strategic Studies* 18 (March 1995), 145–74.

Pearson, Lester Bowles. "Nobel Lecture." 11 December 1957. Available at http://nobelprize.org/peace/laureates/1957/pearson-lecture.html. Accessed 4 March 2007.

Peskett, Gordon R. "Levels of War: A New Canadian Model to Begin the 21st Century." In Allan English et al., eds. *The Operational Art*, 97–128.

Pigeau, Ross and Carol McCann. "Re-conceptualizing Command and Control." *Canadian Military Journal* 3, no. 1 (Spring 2002), 53–63.

Polk, Robert B. "A Critique of the Boyd Theory – Is It Relevant to the Army?" *Defense Analysis* 16, no. 3 (December 2000), 257–76.

Porch, Douglas. "Bugeaud, Galliéni, Lyautey: The Development of French Colonial Warfare." In Paret, ed., *Makers of Modern Strategy*, 408–43.

Pratt, David. "The Way Ahead for Canadian Foreign and Defence Policy." Keynote address at the 20th Annual Conference of Defence Associations Institute Seminar, 26 February 2004, Ottawa, Ont. Available at www.cda-cdai.ca/seminars/2004/pratt.htm. Accessed 16 March 2007.

Pratt, Gerald M. "A Clash of Service Doctrines." In Allan English et al., eds., *The Operational Art*, 225–48.

Pugh, Michael. *Regeneration of War-Torn Societies*. New York: St. Martin's Press, 2000.

Ricks, Thomas E. and Josh White. "Scope of Change in Military Is Ambiguous." *Washington Post*, 1 August 2004, A06.

Roddy, Kimberly A. "Network-Centric Operational Warfare: The New Science of Operational Art." Unpublished US Naval War College research paper, 16 May 2003.

Rousseau, Christian. "Command in a Complex Battlespace." In Allan English, ed., *Leadership and Command and the Operational Art*. Kingston: Canadian Defence Academy Press, 2006, 55–83.

Rudd, David. "Canada's New Defence Policy." The Canadian Institute of Strategic Studies: Commentary, April 2005. Available at www.ciss.ca/Comment_Newpolicy.htm. Accessed 4 March 2007.

Schmitt, John F. "Command and (out of) Control: The Military Implications of Complexity Theory." *Marine Corps Gazette* 82, no. 9 (September 1998), 55–8.

Schneider, James J. *The Theory of Operational Art, Theoretical Paper No. 3*. Fort Leavenworth, Kan.: United States Army Command and General Staff College, 1988.

– "Transforming Advanced Military Education for the 21st Century." *Army* 55, no. 1 (January 2005), 15–22.

"Security Issues Dampen Afghan Investments." *The Associated Press*, 13 July 2004.

Segre, Claudio. "Giulio Douhet: Strategist, Theorist, Prophet?" *Journal of Strategic Studies* 15 (September 1992), 351–66.

Sharpe, G.E. and Allan English. *Principles for Change in the Post-Cold War Command and Control of the Canadian Forces*. Kingston, Canadian Forces Leadership Institute, 2002.

"Shock and Awe: The Idea behind the Buzzwords." *Washington Post*, 30
   March 2003, B03.
Shy, John. "Jomini." In Paret, ed., *Makers of Modern Strategy*, 143–85.
Shy, John and Thomas W. Collier. "Revolutionary War." In Paret, ed.,
   Makers of Modern Strategy, 815–62.
Simpkin, Richard E. *Race to the Swift: Thoughts on Twenty-First
   Century Warfare*. London: Brassey's Defence Publishers, 1985;
   paperback reprint, 2000.
Sloan, Elinor. *The Revolution in Military Affairs*. Montreal and
   Kingston: McGill-Queen's University Press, 2002.
– "The United States and the Revolution in Military Affairs." Ottawa:
   National Defence Headquarters, Directorate of Strategic Analysis,
   Project Report no. 9801, February 1998.
Soeters, J. and R. Recht, "Culture and Discipline in Military Academies:
   An International Comparison." *Journal of Political and Military
   Sociology* 26, no. 2 (Winter 1998), 169–89.
Starry, Donn A. "A Perspective on American Military Thought." *Military
   Review* 69 (July 1989), 6–10.
Stavridis, Jim. "They Got Game." US *Naval Institute Proceedings* 125,
   no. 6 (June 1999), 51–4.
Strand, Arne, Kristian Berg Harpviken, and Astri Suhrke. "Afghanistan
   Peacebuilding in a Regional Perspective." Report of a conference held
   by the Christian Michelsen Institute in Solstrand, Norway, 22–24
   September 2002. Available at www.cmi.no/pdf/?file=/afghanistan/doc/
   afghanistan_conference_report_sept_2002_arne_strand.pdf. Accessed
   5 March 2007.
Strange, Joe. *Centers of Gravity and Critical Vulnerabilities: Building on
   the Clausewitzian Foundation So That We Can All Speak the Same
   Language*, 2nd ed. Quantico, Va.: Defense Automated Printing Service
   Center, 1996.
Sumida, Jon Tetsuro. *Inventing Grand Strategy and Teaching Command:
   The Classic Works of Alfred Thayer Mahan Reconsidered*. Baltimore,
   Md.: Johns Hopkins, 1997.
Swain, Richard M. *Neither War nor Not War, Army Command in
   Europe During the Time of Peace Operations: Tasks Confronting
   USAREUR Commanders, 1994–2000*. Carlisle, Pa.: United States Army
   War College, Strategic Studies Institute, May 2003.
Sweetman, Jack, ed., *The Great Admirals: Command at Sea, 1587–1945*.
   Annapolis, Md.: US Naval Institute Press, 1997.

Swettenham, J.A. *Some Impressions of UNEF, 1957 to 1958: Report No. 78 Historical Section (GS) Army Headquarters*. Ottawa: Department of National Defence, 2 January 1959.

Szafranski, Richard. "Peer Competitors, the RMA and New Concepts: Some Questions." *Naval War College Review* 49, no. 2 (Spring 1996), 113–19.

Thomson, Michael H. and Barbara D. Adams. "Network Enabled Operations: A Canadian Perspective," Human*systems* Incorporated, report written for Defence Research and Development Canada, nd (2005).

Toffler, Alvin. *The Third Wave*. Toronto: Bantam Books, 1981.

Toffler, Alvin and Heidi Toffler. *War and Anti-War*. New York: Warner Books, 1993.

Tritten, James J. "Naval Doctrine... From the Sea." Norfolk, Va.: Naval Doctrine Command, December 1994.

Ullman, Harlan and James Wade. *Shock and Awe, Achieving Rapid Dominance*. Washington, D.C.: National Defense University Press, 1996.

United Nations Organization on Drugs and Crime and the Government of Afghanistan Counter Narcotics Directorate. *Afghanistan Opium Survey 2003*, October 2003. Available at www.unodc.org/pdf/afg/afghanistan_opium_survey_2003.pdf. Accessed 4 March 2007.

US Air Force. *Air Force Basic Doctrine*, Air Force Doctrine Document 1, September 1997.

US Army. *Operations*, Field Manual 3-0, 14 June 2001.

US Department of Defense, Office of Force Transformation. *The Implementation of Network-Centric Warfare*, 5 January 2005. Available at www.oft.osd.mil/library/library_files/document_387_NCW_Book_ LowRes.pdf. Accessed 4 March 2007.

US Navy. "FORCEnet: A Functional Concept for the 21st Century," nd (2005?). Available at www.nwdc.navy.mil/Conops/Files/FnFuncCon.pdf. Accessed 4 March 2007.

– *Multinational Maritime Operations Doctrine Manual*. US Naval Doctrine Command, 1999.

Vallance, Andrew G.B. *The Air Weapon: Doctrines of Air Power Strategy and Operational Art*. London: Macmillan, 1996.

Van Creveld, Martin. *Command in War*. Cambridge: Harvard University Press, 1985.

– "High Technology and the Transformation of War, Part 2." *Journal of the Royal United Services Institute*. 137, no. 6 (December 1992), 61–4.

Warden, John A. *The Air Campaign: Planning for Combat*. Washington, D.C.: Pergammon-Brassey's, 1989.

Warner, Oliver. *Command at Sea: Great Fighting Admirals from Hawke to Nimitz*. New York: St Martin's, 1976.

Wattie, Chris "Absent-minded Professor Didn't Expect a Nobel." *Kingston Whig-Standard*, 13 October 1994, 9.

Westrop, John. "Aerospace Doctrine Study." Unpublished report, 30 April 2002, copy at the Canadian Forces College library.

Wilson, G.I. et al. "4GW – Tactics of the Weak Confound the Strong." *Military.com*, 8 September 2003. Available at www.military.com/ NewContent/0,13190,Wilson_090803,00.html. Accessed 4 March 2007.

Winton, Harold R. "A Black Hole in the Wild Blue Yonder: The Need for a Comprehensive Theory of Air Power." *Air Power History* 39 (Winter 1992), 32–42.

Wyly, Michael Duncan. "Teaching Maneuver Warfare." In Richard D. Hooker Jr, ed., *Maneuver Warfare: An Anthology*. Novato, Calif.: Presido Press, 1993.

Young, Richard J. "Clausewitz and His Influence on US and Canadian Military Doctrine." In Allan D. English, ed., *The Changing Face of War*, 9–21.

Zinni, Anthony. "How Do We Overhaul the Nation's Defense to Win the Next War?" Speech given at Arlington, Va., 4 September 2003. Available at U.S. Naval Institute and Marine Corps Association website www.mca-usniforum.org/forum03zinni.htm. Accessed 4 March 2007.

Index